OSPREY AUTO CHAMPIONS

# CAN-AM CARS

DAVID McKINNEY

OSPREY AUTO CHAMPIONS

# CAN-AM CARS

DAVID McKINNEY

OSPREY
AUTOMOTIVE

First published in Great Britain in 1999
by Osprey Publishing, Elms Court, Chapel Way
Botley, Oxford, OX2 9LP



ISBN 185532 900 X

99 00 01 02 03 10 9 8 7 6 5 4 3 2 1

Editor Shaun Barrington
Page design Paul Kime, the Black Spot
Artwork profiles by David Russell

Origination by Grasmere Digital Imaging, Leeds, UK
Printed in China through World Print Ltd.

## PICTURE ACKNOWLEDGEMENTS

Mirco Decet: pp 39(t), 52-53, 83, 95.

Jaguar UK: p 8(t).

LAT Photographic: pp 6, 7, 8 11, 12, 14-18, 20-24, 25(t), 26-34, 36, 38, 39(b), 49, 42, 44, 46, 47, 50, 54, 55-57, 59, 61, 76, 70(b), 79 80(b), 81, 93(b), 94(bl), 98-104, 106, 107, 109.

Porsche AG, Stuttgart: back cover, pp 69®, 71-77, 82, 90, 92(b), 94(br), 96.

Gridwork/Sutton Photographic: p 94(t).

William Taylor: pp 1, 3, 86, 89, 105.

Michael Turner: front cover, pp 10, 25(b) 45, 48, 60, 64-68, 69(t), 70(t), 80(t), 84, 91, 93.

With special thanks to Mirco Decet for picture research, to Jens Torner, Porsche and to William Taylor, author and photographer of the highly recommended *The Lotus Book: The Complete History of Lotus Cars*.



**The Marketing Manager, Osprey Direct UK, PO Box 140, Wellingborough, Northants, NN8 4ZA, United Kingdom. Tel. (0)1933 443863, fax (0)1933 443849. Email: info@OspreyDirect.co.uk**

**The Marketing Manager, Osprey Direct USA, PO Box 130, Sterling Heights, MI 48311-0130, USA. Tel. 810 795 2763, fax 810 795 4266. Email: info@OspreyDirectUSA.com**

Visit Osprey's website at:
**http://www.osprey-publishing.co.uk**

Half-title page **McLarens, starting with the M1B model pictured, were the most successful cars in the early years of the Can-Am series**

Title page **Another early McLaren, the M1C. Many were raced by private entrants in the early years of the Championship, frequently picking up more than respectable placings behind the newer factory entries**

Front cover **Ferrari brought variety, glamour – and different engine notes – to Can-Am, without ever achieving much success. Any two Osprey Automotive titles in print free to the first three persons to identify via e-mail (see left) the driver, the car, the venue and the year depicted in Michael Turner's superb painting**

CAN-AM

# CONTENTS

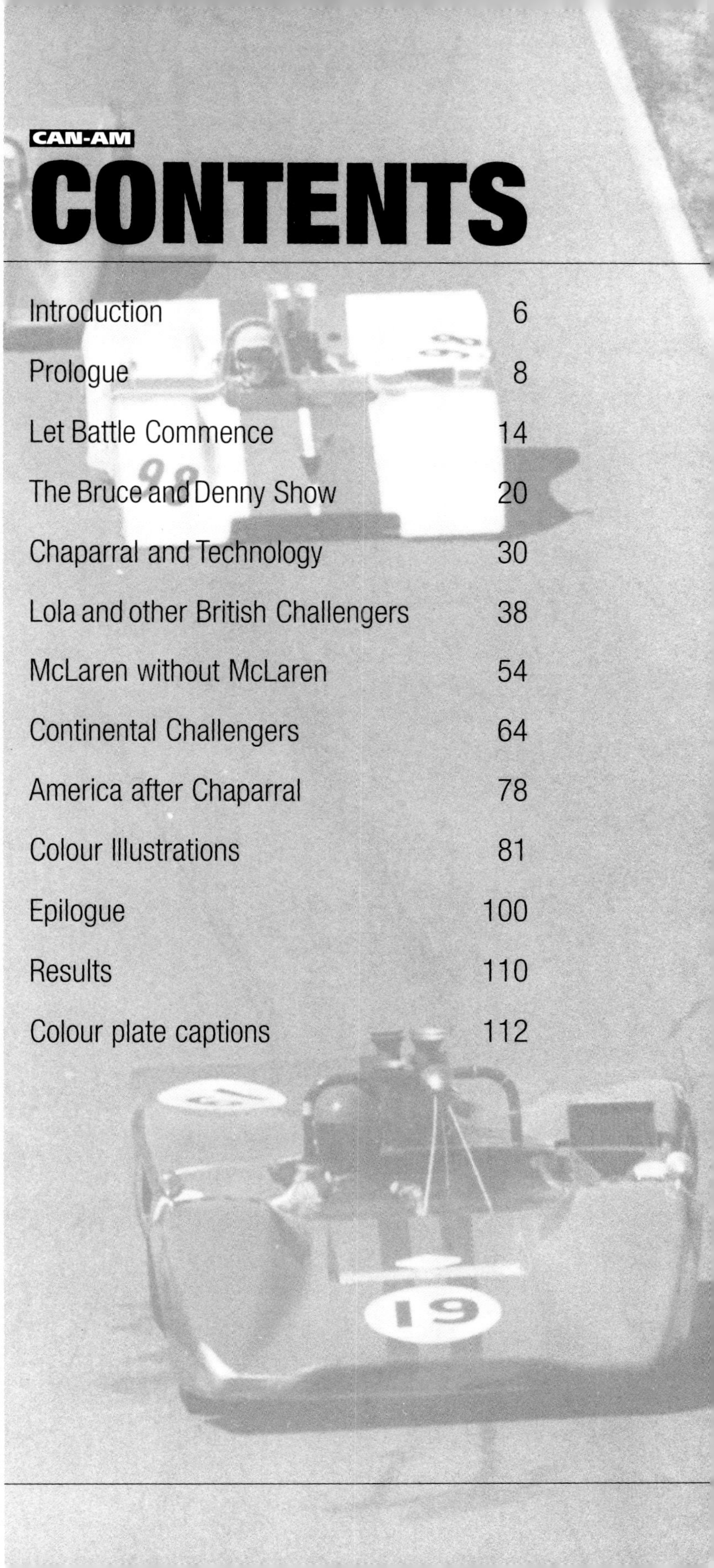


Introduction 6

Prologue 8

Let Battle Commence 14

The Bruce and Denny Show 20

Chaparral and Technology 30

Lola and other British Challengers 38

McLaren without McLaren 54

Continental Challengers 64

America after Chaparral 78

Colour Illustrations 81

Epilogue 100

Results 110

Colour plate captions 112

# CAN-AM

The very words bring to mind a spectacle of colour and noise, and conjure up images of packed fields of big brutal sportscars thundering around North America's racing circuits.In its original form, the Canadian-American Challenge Cup was disputed in the United States and Canada for only nine years in the 1960s and 1970s, yet was a very important part of the international motor racing calendar of the day.

**Jackie Stewart in the Chaparral 2J 'sucker car' chases Peter Revson in a Lola T220 at Watkins Glen in 1970. Both drivers were F1 regulars, two of the many who contested the series**

**Series regular Lothar Motschenbacher in a McLaren M6B leads the modified Lola T160 of John Surtees at Bridgehampton in 1968. The Lola's high wing was one of the first seen in the series**

THESE WERE THE MOST POWERFUL cars of their era, for the series was – deliberately – organised with no upper limit on engine size. Nor for that matter a specified minimum weight. Let the Europeans fuss over the size of Formula One engines, the North Americans said, or how high the windscreen of a sportscar should be. In our series you can race whatever you like. The only rules we believe in are ones that make the car safe.

This approach, so alien to European thinking, gave rise to some weird and wonderful devices as, unfettered by a rule-book, designers experimented with new ways of making their Can-Am cars faster. Nowhere was this more successful than in the field of aerodynamics. Until this time the term, in the motor racing context, was no more than a synonym for streamlining. But it was in the Can-Am series that aerofoils or 'wings' were first used in international racing. The Can-Am was also the birthplace of 'ground effect' technology. Both translated to Formula One and permanently changed its face, and that of every other international category.

Both developments, incidentally, sprang from the fertile mind of Texan Jim Hall, whose Chaparral cars more than any other epitomised the departure from accepted design practice which the Can-Am series not only allowed, but actively encouraged.

Throughout the nine-year history of the series, just about every successful car used an American V8 engine, in most cases Chevrolets. Where the competition element came in, and made the series truly international, was in the cars. McLaren and Lola from Britain were matched against Chaparral and later Shadow from America, with famous Continental marques

Ferrari and Porsche also playing major parts from time to time.

It was a drivers' championship, but the best driver in the world has never won much in the worst car, so cars were always at least as important. Lola won the inaugural title in 1966, but then McLaren launched into a domination which would last for the next five years. In each of those years a works McLaren driver won the title, and in every instance except one his team-mate was championship runner-up. The opposition tried hard, but it was not until the advent of turbocharger technology and massive development budgets that Porsche knocked McLaren from their perch.

Also contributing to the international appeal of the Can-Am series were the drivers. The Formula One stars of those days did not restrict their appearance to Grands Prix, and many of the world's top names were regular competitors in the Can-Am series: world champions Brabham, Stewart, Phil Hill, Graham Hill, Surtees and Hulme were all seen in action in the North American races, as were other F1 regulars of the time such as McLaren, Amon, Gurney, Bonnier, Rodriguez, Siffert, Cevert and Spence.

But further competitive interest was added on the driving front from the American side, not only Phil Hill and Gurney, who were as well known in Europe as in their home country, but also Jim Hall, Mark Donohue, George Follmer and other stars of domestic sportscar racing, and heroes of the Indycar circuit such as Mario Andretti (a future world champion in F1), A J Foyt and Parnelli Jones.

Gradually however, the biggest names became increasingly unable to find the time for pursuits other than those connected directly to their principal employment. One by one the manufacturers that had supported Can-Am pulled out. By the end of its life the once-glorious championship was little more than just another domestic North American series. It was finally killed off by the imposition of more and more rules, the very absence of which had ensured its original success.

This book is not a race-by-race review of the series, but rather an attempt to tell its story by describing some of the most successful and more interesting cars which took part in its nine-year history, and analyse some of their most significant races. If, along the way, we recapture some of the excitement that was the essence of the Can-Am, we will have succeeded in our task.

**Can-Am popularity with the paying public was reflected in huge crowds, here watching the Porsche 917/10s leading the McLarens at Laguna Seca in 1972**

CHAPTER ONE **CAN-AM**

# PROLOGUE

**The Can-Am series was a merging of two great sportscar racing traditions, the European – especially British – and the North American. In Europe the biggest sportscar races in the 1950s were still the classic long-distance events such as the Le Mans 24 Hours, the Mille Miglia and the Targa Florio, all of which dated back at least as far as the 1920s. Britain, in common with most countries, also ran national events, for the same sort of car.**

THE AMERICAN TRADITION WAS much newer, having sprung from the postwar sportscar-buying boom. Like their counterparts on the other side of the Atlantic, US enthusiasts raced Ferraris, Maseratis, Jaguars, Aston Martins, MGs and Porsches.

Again in common with Old World countries, there was a national sportscar championship in the United States. A major difference however, and one which contributed directly to the birth of the Can-Am, was that sportscars were the main attraction at national road-race meetings. In Britain and in Continental countries, sportscar races were not considered as important as events for single-seater racing cars, especially Formula One. Of course, America did have single-seater racing, but this was restricted to relatively short oval tracks, ranging from quarter-mile dirt-surfaced speedways through paved mile ovals to the longer superspeedways epitomised by Indianapolis, the most famous of them all. On road courses, the sportscars were the attraction.

**By the 1960s cars for non-championship racing had started to develop along different lines to endurance contenders, such as the descendants of the C-type Jaguar, here winning the 1953 Le Mans 24hrs**

Native ingenuity and a willingness to try different solutions to engineering problems were two attributes which would later epitomise American designers' approach to the Can-Am. They were already in evidence in US sportscar racing in the 1950s, especially on the West Coast. California had a long history of hot-rodding, and produced a number of limited-production sportscars and one-off specials which, powered by a variety of modified production V8 engines, could often give even the best of the imports a run for their money.

The cars which contested British national sportscar races remained the same as those seen at Le Mans and elsewhere until a 3-litre engine-limit was introduced for the world championship in 1958. The life of such cars as the 3.8-litre D-type Jaguars and Lister-Jaguars was however extended

**Racing developments of American V8 engines, (this one a 1968 version) grew out of the the 1950s habit of bolting bigger local engines into European Ferrari, Jaguar or Cooper chassis**

because British races were not run to world championship rules.

By this time a lot of American drivers had tried racing in Europe, and Masten Gregory, Carroll Shelby and Phil Hill had all made names for themselves on the great road circuits of the world. Their experience at this level in turn raised the game back home when they returned to race at the end of each year. Inevitably, people started to think it would be nice if the US could stage races in which the roles could be reversed, and the Americans could play host to European stars. It already happened in the Sebring 12 Hours each March, but before the concept could be extended to other venues, an important hurdle had to be overcome. In Europe, race organisers paid drivers appearance money and prizemoney, and there were also bonuses from fuel, tyre and component manufacturers.

In American sportscar racing circles this was seen as an affront to the amateur tradition in which the sport in the US was grounded, and the Sports Car Club of America insisted its driving members should not receive any financial reward for their efforts. The point was now made that the Hills, Gregorys and Shelbys would clearly attract just as many paying spectators through the gate as any likely European stars who might be invited. The SCCA had been forced to provide dispensation for its members to race against the professionals at Sebring, provided they didn't accept any money. But the club would not countenance a waiver for any other races.

Above **The Lotus 19 became the car to beat in non-championship races on both sides of the Atlantic. This one has a V8 Ford engine in place of the standard Coventry Climax four**

Below **The Lotus 30 of 1964 was also powered by Ford, but could not compete with the first V8 Lolas and McLarens that were making their first appearance at the same time**

**Stirling Moss, who was front man for Can-Am series sponsors Johnson's Wax, here listens to Chris Amon's views on the Ferrari contender at Watkins Glen in 1969**

As the debate continued, the United States Automobile Club, which controlled racing on the ovals, started running its own road races for sportscars. And paying prize money. The SCCA tried to prevent its members taking part, but when USAC helped stage a sponsored sportscar spectacular at Riverside in California in October 1958, the writing was on the wall. A huge crowd turned out to see the event, in which leading players from USAC ovals, the NASCAR stock-car circuit and national sportscar racing were pitted against big international names such as Moss, Behra, Bonnier and Salvadori, not to mention Hill, Gregory and Shelby. In the face of the success of this event, the SCCA could not do other than swallow its pride and relax its policy.

The Riverside race became an annual fixture, and in 1960 was joined by a second Californian race, at Laguna Seca. At the same time Canadian promoters were heading in the same direction, and began staging two annual races at Mosport Park in Ontario. The other principal component in the informal international series which was taking shape was the Nassau Speed Week, which had been run in the Bahamas at the end of each year from 1954.

Among the drivers who regularly took part in these races were some of the biggest names in the world. They included Stirling Moss, 1959 and 1960 world champion Jack Brabham, his sometime team-mate Bruce McLaren and their fellow F1 driver Innes Ireland, as well as European-based Americans Masten Gregory and Dan Gurney.

By the early 1960s, the leading cars in the big races in California, Canada and the Bahamas were no longer the Ferraris, Aston Martins and Porsches which were disputing Le Mans, Sebring and the other world championship races. Instead the American races were dominated by British Cooper Monacos and Lotus 19s, which had rear-mounted 2.5-litre Coventry Climax engines as standard wear. Both were built with the popular British sprint events in mind rather than the longer-distance championship races, and would find a ready market among the US sportscar-racing fraternity. Moss, Gurney and Gregory all won races in Lotus 19s, while Brabham and McLaren provided the back-up with Cooper Monacos.

## HYBRIDS

With no need to restrict engine-sizes, some of the local entrants were soon fitting small production-based V8 Buick or Ford engines to their Coopers and Lotuses. Roger Penske's infamous Zerex Special retained the four-cylinder Climax unit but strained the regulations in other directions by being no more than a wide-bodied Grand Prix Cooper, complete with central driving position. This was a bit much even for the American regulators, and they soon clarified their definition of a sportscar to make sure that the two mandatory seating positions were on either side of a car's centre-line. But that was not before Penske, in a blaze of international publicity, had won the two big 1962 Californian races.

At almost exactly the same time Carroll Shelby, now retired from race driving, put a small-block Ford V8 engine in an English AC Ace chassis to create the legendary Cobra. The following year he turned to the open class and campaigned a team of Cooper Monacos which used the same Ford engines and were known as King Cobras. Dave MacDonald in one of these won both Californian races in 1963, and USAC star Parnelli Jones won another the following year. The Ford Motor Company itself was behind a V8 Lotus 19 variant which Dan Gurney showed to have great pace, though not the reliability to match.

Now, for the first time, the British and American traditions came together as British manufacturers got into the hybrid act themselves. McLaren won the 1964 Players 200 at Mosport with a Cooper-Oldsmobile, and then in the first McLaren-Oldsmobile was the pace-setter in later North American events that year. The following year both McLaren and his young New Zealand compatriot Chris Amon won races in Canada with McLaren-Oldsmobiles, though they were out of luck in the US races. Lotus meanwhile built their Ford-powered 30 and 40 models, which were not a success. Lola's Chevrolet-powered T70 on the other hand enabled world champion John Surtees to win in North America in 1965.

Right **Typical mix of cars in North American races which preceded the Can-Am. Scarab leading Chaparral, Ford GT40 (a refugee from the world series) and McLaren at Riverside in 1965**

65
7

**The first true McLaren. Creator Bruce McLaren wheels out his Oldsmobile-powered Mk1 for a 1964 race. Built to comply with current sportscar regs, it has lighting and a spare wheel**

None of these cars was aimed at the long-distance classic races, but instead at non-championship events in Britain and elsewhere in Europe. These British manufacturers all looked to sales of their cars as a major part of their income, and were fully conscious of the importance of the North American market.

So, in the mid '60s there were British cars with British engines, and British cars with American engines.

There were also American cars with American engines. Notable among these were the King Cobras already mentioned, together with the Scarabs and Genies, both of which were also built in California and which used a variety of engines, and Jim Hall's Chevrolet-powered Chaparrals from Texas. Hall himself had been taking places in the North American races right from his first front-engined Chaparral in

**Lotus 19 prototype undergoing tests in England. This model, particularly when powered by the 2.5-litre Climax engine, was virtually unbeatable**

1962. With a rear-engined model in 1964 Penske won at Laguna Seca and, with Hap Sharp, was first in the Nassau Trophy. The following year Hall beat McLaren to win the Canadian Grand Prix while Sharp was first in the Riverside race and took a repeat Nassau victory.

As the North American international races grew in importance, USAC had continued to promote professional sportscar races on a national basis, and by 1962 had been looking at formalising a championship. Only now did the SCCA finally accept the inevitability of a professional series, and in 1963 preempted the USAC move by announcing a national pro championship of their own. The United States Road Racing Championship, as it was called, was dominated by Hall and Sharp in Chaparrals in both 1964 and 1965, against the King Cobras and the first Lola T70s to be sold in America. Hall was champion in 1964 but the vagaries of the points-scoring system gave the 1965 title to George Follmer, who drove a Porsche-engined Lotus 23 in the 2-litre class.

There were no Chaparrals in the 1966 series. Their biggest success of 1965 had been victory in the Sebring 12hr classic, and the team concentrated its 1966 efforts on the international endurance championship. They would win world title rounds both that year and the next, but that's another story.

By now most of the leading USRRC players were running Lolas or McLarens, and race-winning drivers in 1966 included several names which would become familiar to followers of the Can-Am series when it got under way later that year. Among them were Mark Donohue, who drove a Lola T70 entered by the now-retired Roger Penske, and German-born Lothar Motschenbacher in a McLaren. Charlie Parsons, who would win the USRRC title this year, and Canadian John Cannon both started the series with Genies, but switched to McLarens before the season was over. The existence side by side of the domestic USRRC championship on the one hand and, on the other, the independent international events at the end of each year was a situation crying out for amalgamation. Their inevitable coming together as an international championship series was formalised with the launch of the Canadian-American Challenge Cup series in 1966.

**Jim Clark doing his best (which was usually pretty good) in a Lotus 30. Something of a lemon, its replacement, the Lotus 40, was even worse, famously described as "a 30 with ten more mistakes"**

CHAPTER TWO CAN-AM

# Let Battle COMMENCE

The Sports Car Club of America and the Canadian Automobile Sport Clubs, as joint promoters of the Can-Am Cup series, ensured support from a wide range of potential competitors by offering prize money at what by the European standards of the day was an astronomical level. An average of $30,000 was at stake in each of the six scheduled races, with an end-of-year fund boosted to $55,000 by series sponsors Johnson Wax. Not only could a successful competitor cover costs, he could actually make money – first prize would pay enough to buy a year-old car. This aspect of the series appealed to the European teams just as much as to the Americans, and several semi-professional British competitors sent in their entry-forms.

**Chapparral's contender for the first Can-Am Championship was based upon the 2C pictured. In 1966 (2E) form the cars were distinguished by revolutionary suspension-mounted aerofoils**

SO TOO DID OFFICIAL or quasi-official teams from McLaren and Lola, attracted not only by the cash, but by the equally attractive prospect of selling cars to private entrants.

The same reasoning was applied by the smaller American car-builders. There were Genies, McKees and Websters in at least the early races, but it quickly became apparent that in this company they were now little better than midfield runners. Only Chaparral of all the American makes was regarded by the two big British constructors as being likely to provide them with serious competition.

McLaren and Lola had shared the spoils between them in earlier races that year, in both Britain and North America. But that was without Chaparral opposition, and Jim Hall was known to have a new contender on the drawing-board. If the Texan cars could win against the British cars in domestic US events in 1964 and 1965, what would the latest car do in 1966?

It seemed most likely that by the end of the series the Can-Am champion would be the driver of a Lola, McLaren or Chaparral.

Lola had started in business building and selling sportscars in 1958, and were one of the first constructors to put an American stock-block V8 engine in the back of a British sportscar. The Lola Mk6 of 1963 is best described as promising rather than successful, but led directly to Eric Broadley, the company founder and chief designer, being invited to join the Ford design team to help create the famous Le Mans winning GT40 coupé. Back at his base, he developed the concept into the open Lola T70, which first appeared in 1965.

McLaren were the new kids on the block, the first car to bear the name having appeared only in late 1964. This could trace its ancestry back to Roger Penske's Zerex Special,

which McLaren had bought, revising the chassis and installing a small-block Oldsmobile V8 engine. After early success with this he applied the lessons learned to the McLaren Mk1.

Jim Hall at Chaparral had a number of advantages over both Lola and McLaren. First and foremost, perhaps, was the relationship he had built up with Chevrolet's Research & Development department, which gave him access to engines and technical assistance not available to anyone else. He also had his workshops sited at a racing circuit, Rattlesnake Raceway, which meant that innovations could be tested as soon as they were developed. Other teams had to hire circuits for testing purposes and transport their cars to them, return to base for modifications, then repeat the whole process.

Without doubt though the biggest ace in Hall's pack was his own design genius, an ability to abandon accepted wisdom and apply lateral thinking to the sort of engineering problems that faced anyone trying to produce a race-winning car.

**Concentrating on running their own team, McLarens contracted the building of customer replicas to the Elva company. This M1B was campaigned in the 1966 series by Lothar Motschenbacher**

During the life of the rear-engined Chaparral 2, the car which had dominated the USRRC in 1964 and 1965, he tried a number of developments, with aluminium-block engines, with semi-automatic 'torque converter' transmissions, and also in the field of aerodynamics. To help push the nose of the car onto the road, and assist its cornering power, a 'spoiler' was fixed below the nose, and the tail-section was swept up to provide similar assistance at the back.

65

Left **Mark Donohue in Roger Penske's beautifully presented Lola T70 leads the high-winged Chaparral 2Es of Hall and Hill on his way to winning the Mosport round**

These devices were just starting to be used in sportscar racing in Europe, but the all-new Chaparral 2C, which appeared briefly at the end of 1965, broke completely new ground. Its rear spoiler, mounted across the car between finned rear wings, could be adjusted by the driver on the move. You raised it for the twisty bits, to push the car down onto the road and increase its cornering speed, then kept it flat to minimise wind resistance on the straights.

Aerodynamic aids aside, all the main players in the 1966 Can-Am followed similar general design concepts. In those days, the way to build a race car was to start with a box-shaped tubular steel frame for the chassis, braced at strategic places for added strength and to take the loads generated by forward and lateral movement. Or you formed a similar structure by the more modern method of using lightweight sheet-metal boxes.

Whether you used the 'space-frame' or 'monocoque' approach, you then attached wishbones at each corner, perhaps assisted by radius arms, to take the wheels, and coil springs to absorb the bumps and to control cornering forces. The engine and transmission were bolted into the rearmost bay of your box and steering gear, pedals and radiators in the front, leaving the middle space for the driver (and nominal passenger). The whole lot was then clothed in a full-width body made of either aluminium or fibreglass.

## SAME QUESTIONS, DIFFERENT ANSWERS

The 1966 Chaparral and Lola were monocoques, but the McLaren was built around a space-frame. After its initial season the Mk1 McLaren had been revised by aerospace engineer Robin Herd, who had just joined the small company. The later frame was claimed to be 20% more rigid than the original, but with no gain in weight. Its round and square mild steel tubes were stiffened by sheet-alloy panels forming the undertray and bulkheads, with suspension by the standard system of coil springs and wishbones.

Lola construction was semi-monocoque, of sheet steel and alloy, with weight saved by bonding and riveting instead of spot-welding. The car was longer and wider than the McLaren, but handled better, which brought it immediate success against its rivals in the races leading up to the Can-Am. Before the international series the company produced a Mk2 version, with the centre monocoque made entirely of alloy, and therefore lighter than the Mk1.

Chaparral's Can-Am contender was the 2E, built on a steel monocoque with conventional suspension layout similar to its rivals'.

The official teams of all three constructors relied on Chevrolet's production-based V8 engines, though some private McLaren and Lola owners opted for other units. McLaren themselves had first used the comparatively light 3.9-litre Oldsmobile V8, which produced around 310bhp, moving quickly to a 4.5 version which gave a bit more. The 4.7-litre Ford of about 375bhp was also a popular choice with buyers while others opted for the 6-litre Chevrolet. This could give 500bhp, but was bulkier than the rival units, and the amount of iron used in its construction meant it weighed a lot more. Before the Can-Am the McLaren factory switched to medium-sized 5.4-litre Chevrolet engines, which they tuned to give more than 450bhp, though during the season they would experiment with fuel-injection, and with 6-litre engines, and gain about 100bhp in the process. The most serious Lola entrants used 6-litre Chevy engines, while Hall's connections with Chevrolet R&D enabled him to use 5.4-litre aluminium-block engines which were only about two-thirds the weight of the opposition's conventional versions.

The three main teams also had different approaches to transmitting their power to the road. Chaparral stuck with their 'torque converter' system while Lola used a Hewland HD transaxle and McLaren a ZF unit, not because it was necessarily a better design, but because it was lighter.

Where the real differences came though was in the bodywork. The Lola's was designed by John Frayling, who had been responsible for the original Lotus Elite, and was made of fibreglass, while the McLaren's body was styled by the motor racing artist Michael Turner. Both constructors had noted the Chaparral's aerodynamic devices the previous year, and both appeared for the Can-Am with spoilers fixed to the front and back to improve aerodynamic efficiency.

But competitors, officials and spectators alike were all amazed when the new Chaparral 2E was seen in public for the first time in the 1966 Can-Am. Standing high over the rear of the white car was what appeared to all intents and purposes to be an aircraft wing, though closer inspection showed it to be upside down. Hall had been concerned that the increased downforce of the 2C's 'flipper' had necessitated heavier springs, which in turn had adversely affected handling. He rethought the whole problem, and this was his answer. His reasoning was that if an aircraft wing was shaped to provide uplift, then turning it upside down would help glue the car to the road, and so improve cornering. Basic stuff today, but pretty revolutionary back then. It is true that one or two others had tried similar approaches before, but none had persevered.

Connected by long struts to to the rear hubs, the Chaparral's revolutionary aerofoil was again controllable from the cockpit, as was the flap mounted inside the nose to balance the rear downforce. Blocking the air-intake like this meant the radiators had to be mounted on the sides of the car's body, between the front and tear wheels – another departure from usual practice that would become widely copied by other designers in years to come.

Hall had two new Chaparral 2Es ready for the first Can-Am series, and hired retired F1 driver Phil Hill, the 1961 world champion, to drive the second. The two official McLaren entries would be driven by McLaren himself and his New Zealand compatriot Chris Amon, who had provided useful back-up in the past year or two. The principle Lola player in the series would be John Surtees, whose Team Surtees was in effect the works development wing and who, in common with most Lola users, relied on Chevrolet power.

Chaparral did not sell cars to the public, but both McLaren and Lola did, and many privately owners of both British makes filed Can-Am entries. Most prominent among these was American hero Dan Gurney, who had recently set up his All American Racers operation to build and race Eagle cars in Formula One. For the Can-Am he acquired a new Lola into which he installed a lightweight 5-litre Ford engine. Englishman Harry Weslake, designer of the Eagle's F1 engine, devised special aluminium cylinder-heads for the Ford engine which gave it almost 500bhp, and by the end of the season that figure would be up to 520bhp.

There were strong entries too from the USRRC regulars, led by Roger Penske's Lola-Chevrolet which would once again be driven by Mark Donohue, perhaps the finest driver in domestic American road racing at that time.

## THE FORD AND CHEVY SHOW

McLaren had meanwhile completed a deal with Elva Cars, owned by the Lambretta-Trojan Group, to produce cars for sale to private owners, the 1966 model being marketed as the McLaren-Elva Mk2. Newly crowned USRR champion Chuck Parsons led the McLaren privateers with a Chevrolet-powered car, while up-and-coming Lothar Motschenbacher started with a 5-litre Oldsmobile before switching to a Chevy of his own.

The 1966 Canadian-American Challenge Cup series was run over four races in the United States and two in Canada, each of around 200 miles. Sprint races they weren't, but neither did they approach the duration of the 1000km or 24hr races which still tended to make up the world championship. There were big entries for every round, and even though many runners could hardly be described as competitive, their presence ensured that full grids of 30 or 35 cars started every race. The North American paying public loved the Can-Am races, and attendance records were broken all over the continent.

The first round was held at Mont Tremblant-St Jovite in Quebec. Thirty-four cars qualified for the race, all but five using either Chevrolet or Ford engines. Most were either McLarens and Lolas, which between them set the fastest 19 times. Chaparral encountered transport problems on the way to Canada and had to miss the race.

**F1 stars John Surtees and Pedro Rodriguez discuss their chances at Bridgehampton Can-Am round in 1966. Surtees would win that year's series, but Rodriguez' small Ferrari retired here**

The series opener was between Surtees in his Lola and the two works McLarens. The Lola started from pole position on the grid with the McLarens alongside. All three put up a tremendous show, and the spectators left the circuit knowing they had seen the birth of a great new class of motor racing. McLaren had a turn in the lead but Surtees got ahead again and, after an exciting battle with his rival, eased away to win by 6.5 seconds. The other works McLaren lost two laps early on having a sticking throttle repaired, but Amon fought his way back to third at the end.

But the Chaparrals would probably have been faster. When the white car joined the series for the Bridgehampton race in New York state, Hall qualified fastest of all. The other team car crashed however when its wing collapsed, and when Hill took out Hall's car much the same thing happened. One car was repaired for Hill to drive in the race but after a strong early showing it struck more wing problems and dropped to fourth at the end. Gurney's AAR Lola-Ford had been a non-starter in the first race when his engine expired in the warm-up, but made up for it by leading the Bridgehampton race from start to finish. The McLaren team had now fitted their 6-litre Chevrolet engines, and again Amon was the star, closing on Gurney in the late stages of the race and missing victory by less than a second. McLaren finished third but Surtees retired.

Gurney also started from pole at Mosport, and had built his lead up to more than a lap when, with just ten laps to go, his engine broke its crankshaft. That left the McLarens heading for an historic 1-2 until both were put out by accidents involving other cars, and Donohue in his private Lola was left to win. Hall had again lapped faster than anyone else in qualifying, but his time was set in a non-counting session, and he in fact started from tenth on the grid. The car was a retirement in the race, though Hill brought the other home second. Surtees had not lasted the distance in this race either.

Things finally came right for Chaparral at Laguna Seca, where no-one could match the pace of the high-winged white cars. Hall and Hill set the two fastest times in qualifying, and then Hill won the first half of the two-part race, with Hall second. The Chaparrals again dominated the second race, this time with Hall in front, until he dropped back to let Hill past. They even let a competitor get ahead, for they knew he could not affect the overall positions, and on combined times of the two races Hill and Hall were first and second overall. Gurney was a retirement and so, for the third time in as many races, was Surtees, while the best-placed of the works McLarens, after a fraught race, was fifth.

Hill's win, added to his earlier placings, put him in the championship lead, and American hopes for the remaining races in the series were high.

At Riverside McLaren had fuel-injection fitted to his engine and that helped him to qualify faster than all the opposition, but neither team car made it to the finish. Surtees ran home the winner, though only after Hall was forced to slow with an overheating engine.

Until then the Lola and Chaparral drivers had thrilled spectators with one of the best battles of the season. Team Surtees entered a second Lola here, for 1962 world champion Graham Hill, and he finished third, behind Hall. Hill in the other Chaparral had been an early retirement.

The Chaparrals started the last round, at Las Vegas in Nevada, from the first two places on the grid but wing problems returned to plague Hall. Hill lost time with a variety of problems and eventually finished seventh – out of the points. Gurney retired for the third time in a row and Amon, who had been fastest qualifier after the Chaparrals, also failed to finish. Surtees won the race with McLaren second, and the Lola driver was crowned champion.

## PRIVATEERS

Donohue emerged as the most successful privateer of the series for, after missing the first round, he scored points in all the others. His win at Mosport was backed by a third place, two fourths and a fifth in his other finishes, which left him second on the points table behind Surtees. and ahead of McLaren, Hill, Hall, Amon and Gurney, who took the next places in that order.

Among the many other independent Lola entries, A J Foyt and Mario Andretti appeared at Riverside with a pair of cars powered by 7-litre Ford engines but both retired. Andretti, whose car used automatic transmission, was out again at Las Vegas but again retired, this time on the first lap.

Parnelli Jones, like Foyt and Andretti a star of USAC oval racing, used a supercharged Ford engine, but the only time the car really shone was with a borrowed Chevrolet unit at Laguna Seca. Although starting from 27th place on the grid, Jones scored a brilliant win. That was however only one half of a two-part race, and trouble in the other heat meant he was classified seventh overall. His entrant fielded a second car for Scottish Formula One ace Jackie Stewart in the last three rounds but in spite of showing strongly it did not finish. In an unsupercharged Lola-Ford however former USRRC winner George Follmer took two fifth places during the series.

Apart from those already mentioned, a number of other British-based drivers also contested at least some of the series, all remaining faithful to the Chevrolet engine. Denis Hulme joined in at Mosport with with the Sid Taylor T70 with which he had enjoyed a successful British season, and fought an incredible battle with his countrymen Amon and McLaren before retiring with a broken half-shaft. He also went out at Laguna Seca and Riverside. Paul Hawkins and Hugh Dibley managed five top-ten placings between them, but the best was a fifth by Hawkins.

In a reflection of the overall situation, the private McLaren drivers did not fare as well as their colleagues who had opted for Lola. Parsons placed in the top six in three races, with a third as his best placing, but Motschenbacher did not do quite so well. He did gain points however, as did fellow McLaren men John Cannon and former F1 drivers Peter Revson and Masten Gregory. Cannon used a Ford engine and fellow Canadian Ludwig Heimrath tried a 7-litre Ford in his car, but it was too heavy.

CHAPTER THREE **CAN-AM**

# The Bruce and DENNY SHOW

The first Can-Am series had been a resounding success, and the three main contending teams returned to their respective bases to work on improved contenders for 1967. Lola and Chaparral would build better and better cars each year, and would remain strong contenders. There were new challengers too. But none could match the mighty McLarens, which would be virtually unbeatable for the next five seasons. There were 43 races in that period, and works McLarens won 37 of them.

**Denis Hulme, in cockpit, confers with team leader Bruce McLaren at one of the 1969 rounds. Between them, the New Zealanders won 20 of the 23 races held over the three seasons to 1969**

THE THREE YEARS FROM 1967 to 1969 were the pinnacle. Each year McLaren came up with an improved car, and each year ran the opposition into the ground. Bruce McLaren himself was champion the first year, by a small margin of points from his new team-mate Denny Hulme. In 1968 Hulme took the title, with the boss second. And it was McLaren's turn again the following season, with Hulme a dutiful runner-up. Not for nothing did the Can-Am series become known in this period as the 'Bruce and Denny Show'.

Following the first series, Robin Herd had set about devising an all-new car to outperform the Lolas and Chaparrals in 1967. Assisted by Gordon Coppuck, who had joined the firm as draughtsman the previous year, he started with a clean sheet of paper and devised a full monocoque. This was made of bonded and riveted sheets of magnesium and aluminium, welded to square-section steel tube, with fuel-bags in both side pods and under the driver's knees. Suspension was on the same principle as the earlier models, but McLaren were now making their own wheels, which were fitted to the new car, and carried 13.5 inch tyres. The steady increase in tyre widths would be a feature of the series over the years, the better to take advantage of the continuing increase in aerodynamic downforce.

In the design of the M6A, as the 1967 monocoque McLaren was called, Herd paid particular attention to aerodynamics. The fibreglass body was wedge-shaped at the front for increased downforce, and tested for effectiveness in a wind-tunnel. A Chaparral-type suspension-mounted rear aerofoil was tried as part of the exercise, but discarded, and by the time the car appeared in public it had a fixed spoiler across the tail.

New engine man Gary Knutson, who had been hired from Chaparral, called on his contacts in the Californian hot-rod industry to acquire the necessary bits to assemble special 6-litre Chevrolet engines which, using fuel-injection, produced 525bhp. These were mated to Hewland's new five-speed LG600 gearbox, a lighter unit than the old HD.

Perhaps the most noticeable change between the 1966 and 1967 cars however, was the colour: the M6As were presented in the pale orange or papaya colours which would become a McLaren trademark.

Significantly, the cars were ready more than two months before the first race, and the time was used for an intensive on-track development programme at Goodwood, Snetterton and Silverstone. The already excellent design was thus fine-tuned to the point where the car was one of the best handling in the history of the Can-Am series.

Chris Amon had now signed with Ferrari for Grand Prix and sportscar racing, but McLaren found another of his countrymen, Denis Hulme, as a replacement. One of the fastest Lola drivers in 1966, Hulme was on his way to winning the 1967 Grand Prix world championship with the Brabham team.

This year's customer car was the M1C, and sales of these helped contribute to the fact that, of the 32 cars which started the first round of the 1967 Can-Am series, on the Road America circuit at Elkhart Lake in Wisconsin, no fewer than seventeen were McLarens.

This race started the pattern of domination by the two McLarens which would continue all season – and beyond. McLaren qualified on pole with Hulme second, and although the team-leader went out of the race when his engine ran its bearings, the new signing cruised home to victory. At Bridgehampton Hulme started from pole with McLaren alongside, and they finished the race in the same order. The grid order was then repeated at Mosport, where Hulme completed his hat-trick of wins, though not without leaving the road. This incident, less than two laps from the end, damaged the bodywork which in turn slashed a front tyre, and the McLaren completed the race laying a dense – and smelly – smoke-screen. McLaren, in spite of losing

**McLaren in his brand new M6A at Elkhart Lake for the opening round of the 1967 season. Helping to keep an eye on the instruments is long-time team engineer Tyler Alexander**

Top **McLaren M8A of 1968 was designed to take the newly-available aluminium-block 7-litre Chevrolet V8. Ever bigger engines were a continuing trend throughout the life of the series**

Above **Four-speed Hewland LG500 gearbox was used by 1968 McLaren to save weight over the five-speed version. Engine was a stressed part of the overall chassis design**

**Towering trumpets for the fuel-injection system were a feature of the 1968 McLaren. We believe that's rival Roger Penske (right) lending a hand and no doubt making a careful study - can anyone out there confirm?**

a minute at the start while a leaking fuel-cell was changed, again finished second. To rub salt into the opposition's wounds, F1 driver Mike Spence, one of Chaparral's drivers in the world championship, was third in a privately-entered 1965-model McLaren.

Over the second half of the season the Lola and Chaparral opposition began to catch up, and McLaren was beaten for pole position at Riverside. He was fastest qualifier at both other rounds however, and his race results – which is all that really counted – were first at Laguna Seca and first at Riverside, though an oil-leak put him out of the Las Vegas round. But Hulme's early run was over, and he retired in all three rounds. On one occasion he was black-flagged because of damaged bodywork, the result of a marker-tyre having been thrown up by another car, and in the other two races his engine blew. But the M6As had been the class of the field all year, and the opposition knew that the New Zealanders had only to keep going to win.

Final championship standings were McLaren first, Hulme second, with the Lola and Chaparral opposition nowhere. Spence had however taken another third and a fifth, to share sixth place on the ladder.

Swiss-born Jo Marquart, a former Lotus designer, was brought in to oversee development of the 1968 McLaren defender, the M8A. Assisted by Coppuck, and with Bruce McLaren as always playing a major part, Marquart followed the M6A concept but designed a simpler monocoque, in which the engine formed a stressed part of the structure. Other constructors had started using weight-saving aluminium or magnesium for the bulkheads which formed the transverse bracing of their monocoques, but the McLaren stuck with stronger steel structures. Magnesium was however used for the floor and in other places.

Some had begun using bigger engines than the McLaren's and the M8A was designed larger than its predecessors, and with more fuel capacity, so advantage could be taken of this trend. The size and weight of the freely-available 7-litre Chevrolet engine had dissuaded most contenders from using it in the past, but versions were now being produced with light-alloy blocks, and it was Traco-modified versions of these that McLaren would use in 1968. Further developed by Gary Knutson and Lee Muir in McLaren's new Los Angeles engine shop, the engines produced 620bhp, some 100bhp up on the 1967 figure. But the gain was at the cost of reliability, and a lot of problems had to be overcome before the 1968 McLaren-Chevy could be relied on to produce the sort of results the 1967 model had.

Above **The famous wet Can-Am round at Laguna Seca wasn't the only 1968 race affected by rain. Here team boss Roger Penske shows his driver Mark Donohue his advantage in a pre-season USRRC race**

Right **Donohue again, this time chasing Hulme at Laguna Seca. Both were outrun by Canadian John Cannon in an older McLaren, but fitted with more suitable tyres**

The bigger car meant that weight-saving was important, and a reduction was made by the use of the four-speed LG500 Hewland gearbox. In any case, it was argued, with no standing-starts in Can-Am races, there was little use for five speeds. The suspension layout was redesigned at the same time, bigger brakes fitted to stop it all, and the whole lot clothed in a new and smoother body. Overall, the M8A was lower and wider, accentuated by the fact that even wider tyres were now being used. A row of distinctive foot-tall fuel-injection trumpets projected out of the rear deck, with a shallow aerodynamic wing behind. The net result was a package that was no heavier, but 20% more powerful.

McLaren's increasing involvement in Grand Prix racing meant there was a rush to get the Can-Am cars finished in time for the series. But there had been time before that for the prototype to be tried and proved, and the team knew they had another good car. At the same time the previous year's M6A was developed for production and sale by Trojan as the M6B.

In 1967 the works McLarens had qualified fastest for four of the six rounds. This year they went better, and started every race from pole, McLaren four times and Hulme – now a works McLaren driver in F1 as well – the other two. They also started from second on the grid in every round.

Hulme and McLaren finished first and second in the Elkhart Lake series-opener, but both retired with engine trouble at Bridgehampton. The Edmonton round in Alberta brought another Hulme-McLaren 1-2, but at Laguna Seca the works drivers – and just about everyone else – were thrown into confusion when rain turned the track into a skating rink. As the leaders slithered their way around as best they could, attention

Above **Mark Donohue was easily the most successful of the home drivers in the Can-Am. A switch from Lola to McLaren brought him third place in the 1968 championship standings**

Left **Among the cars McLaren had to beat were not just the big two, but a host of local hopefuls, such as the Caldwell, here at Riverside in 1968**

switched to John Cannon, who had obtained some special Firestone 'intermediate' tyres, which enabled his two-year-old McLaren M1B to win. He beat Hulme by a lap while another Canadian, 23-year-old George Eaton, who also had Firestone tyres on his M1C, finished third, two places ahead of McLaren.

Things returned to normal at Riverside where McLaren was winner, but he was delayed by a multi-car first-corner pile-up at Las Vegas, and finished sixth. Hulme had taken fifth place in the earlier of these two races after going off to avoid another competitor, but was the winner in Nevada, and headed the season points score ahead of his team-mate.

There had been little opposition from Lola this year which meant that, with the exception of Hall's Chaparral, most of the back-up was provided by private McLaren entries, the majority of them using the M6B customer model. A surprising number used Ford engines, and among them Dan Gurney's car, built by his Eagle operation, attracted most attention. The car used a five-speed Hewland LG600 gearbox but with modified suspension and body, and incorporating many titanium components, it was immediately dubbed the 'McLeagle'. Among the experiments tried during the season were twin aerofoils, the second in front of the cockpit, but this was abandoned.

In the Lolas he had run in previous years, Gurney had used 6.2-litre Ford engines with special Weslake cylinder heads. He'd sold his 1967 units, only to find he could not buy the blocks he wanted for 1968, so had to fall back on smaller 5.3-litre units, which were modified with Weslake heads and fuel-injection. The car was not a success however. It had had a troubled debut at Bridgehampton, finally limping home sixth, and retired in all its other races.

Another McLaren-Ford project was that of Motschenbacher, who ran an M6B with ex-Gurney 6.2-litre engines and took a third place and a fourth, which left him fifth on the table.

Riverside was one of Hulme's few retirements in 1968, when a slower car forced him off the road. He kept his championship lead though, and clinched it with victory in the next race

Carroll Shelby meanwhile opted for a 7-litre alloy-block Ford engine in his car. In several rounds Peter Revson in this car proved faster than everyone except the works M8As, but by the end of the season had only a fourth place to show for his trouble.

But the most successful of the McLaren independents this year followed the factory lead and used Chevrolet engines in their cars. Best of them all was the much modified car of Roger Penske who, like Gurney, had switched from Lola to McLaren this year, and who also developed his own suspension. The Penske car also had a one-off body, much lighter than standard. Driver Mark Donohue, fresh from winning the USRRC yet again, and also the Trans-Am sedan championship, did not qualify the dark blue Penske McLaren lower than fifth for any round, and was on hand to win at Bridgehampton after the M8As retired. He also took a second place and two thirds to place third on the points table at the end of the series.

Following his famous Laguna Seca victory, Cannon again qualified 15th for the next round, at Riverside, and – in spite of no rain – came home in sixth place, which is also where he ended up on points for the season.

The factory was not idle over the winter, and produced the new M8B model for McLaren and Hulme's 1969 campaign. This was a lighter version of the M8A with refined suspension.

Chaparral, while developing its own cars, ran a McLaren M12 in some 1969 races. With it Surtees, (here at Watkins Glen) was as fast as anyone

Above **The Caldwell again, this time at Las Vegas in 1968. Future champion George Follmer lines up his Lola T70 to pass en route to second place in the race, seventh in the Championship. Without the participation of the privateers and one-offs, Can-Am would have been fatally impoverished**

**Wings were connected directly to M8B rear suspension, mountings visible in this shot. Fuel-injection trumpets were, if anything, even longer than on M8A engine**

Chassis-mounted wings had now been banned from Formula One, but were allowed in the Can-Am for one more year, and the M8B had an adjustable aerofoil mounted high over the rear, and connected by struts direct to the suspension. Outwardly the car was further distinguished from the M8A by cutaway front wings. The engine shop had now been moved back to England, where another American, George Bolthoff, took over its running from Knutson. Some of the existing 7-litre engines were used, but Bolthoff put most of his effort into revising the cylinder dimensions to give a wider rev range and more torque. Power output was given as 635bhp, but in fact more than 675bhp was available. Testing of the new car started in January and, once again, the team arrived in confident mood.

The SCCA had now abandoned its domestic USRRC series, but as compensation the Can-Am was expanded from six rounds to eleven. Circuits not used in the Can-Am before were Watkins Glen (New York), Michigan, Mid-Ohio at Lexington and the combined road/track course at College Station in Texas, which replaced the traditional Las Vegas round as the season finale. All this made it a long, tough series, but McLaren were equal to the challenge. No doubt they were spurred on by the fact that the prize fund, which had been increased every year, now topped one million dollars. The works M8Bs qualified on pole in all eleven rounds, and were also second-fastest qualifiers at every circuit except one. Hulme showed up better this time, taking seven poles to McLaren's four.

Left **The 1969 M8B differed little from the 1968 M8A, apart from the obvious addition of the high suspension-mounted wing. Hulme on his way to winning the Riverside round**

The race results were just as demoralising for the opposition, for McLaren won six rounds – at Mosport, Watkins Glen, Elkhart Lake, Michigan, Laguna Seca and Texas – and Hulme was first in all the other races: St Jovite, Edmonton, Mid-Ohio, Bridgehampton and Riverside. On eight of those occasions the team-mates finished first and second, often staging near photo-finishes, and with Hulme slowing on more than one occasion to let the boss win as per pre-arranged plan. Final points were McLaren 165, Hulme 160, with the third-placed man accumulating only 81. The team trumpeted in their supremacy even louder by running the spare car for Gurney at Michigan and, although starting from the back of the grid, he came through to help the orange cars to the first three places at the end.

Chris Amon rejoined the team for Laguna Seca on a one-off basis, also starting from the back. He got up to eighth, dropped to the tail again with a pit-stop, and was back up to sixth place when his diff failed just nine laps from the end.

But in 24 starts the works cars retired only four times.

Among several drivers who drove the latest Trojan customer model, the M12, was 1966 Can-Am Champ John Surtees, who had now severed his connections with Lola and signed to drive the latest Chaparral. The American car required further development however, and the McLaren was acquired as an interim measure. Fitted with a high wing and one of Chaparral's very special aluminium 7-litre Chevrolet engines, the car battled with the works M8Bs in three rounds, and took one third place. At Michigan Surtees was ill, so the car was driven by Italian Andrea de Adamich from the Surtees F5000 team, and came to the finish-line in fifth place. McLaren's US agent Motschenbacher was the driver at Laguna Seca, but fuel problems left him eighth.

Until then Motschenbacher had campaigned his own M12 with 7-litre alloy-block Chevy, and took four top-six placings. Rising Canadian star Eaton in a similar car took a third, a fourth and a sixth, and then ended the season with second place to McLaren in Texas. This left him fifth in the standings, and the best-placed McLaren privateer.

There were also two important McLaren-Ford projects this year. Gurney started off using a 5.6-litre alloy-block motor with Weslake heads. This 540bhp unit was good enough to keep the 'McLeagle' up with the works cars at Mosport, until a hub-carrier broke, and then he lost his only engine in practice for the St Jovite round. He missed several rounds waiting for a new 8.1-litre Ford but it never eventuated, and he turned instead to a 6.5-litre model, which was tested with a special three-valve head.

Indianapolis 500 winner Mario Andretti joined the series after two rounds with a McLaren M6B entered by stock-car legends Holman and Moody, which used an 8.1-litre alloy-block Ford engine for which no less than 720bhp was quoted. He qualified third at Elkhart Lake behind the inevitable works duo, but broke a half-shaft during the warm-up and was a non-starter. He did however take a third place and a fourth in later races, and in the last round split the works cars in practice – the only time that happened all year. He led that race from the start but was soon second and then after only ten laps went out with a blown engine.

In spite of a busy USAC schedule, Gurney had returned for a handful of outings. He had now given up on Ford engines and his car had a 7-litre Chevy, providing a single fourth place.

Left **Former Can-Am champ Surtees abandoned McLaren when Chaparral's 2H arrived later in '69. The Texan car was not as competitive**

Below **Holman & Moody fielded a customer-model M6B for Mario Andretti in 1969, with modifications that included a strutted wing and a brutal 8.1-litre (429 cubic inch) Ford engine**

# CHAPARRAL and TECHNOLOGY

McLaren might have been the dominant force in Can-Am, but the car that set all the trends each year – and the one the New Zealanders feared most – was the Chaparral. Having proved in 1966 that the 2E was a winner, Jim Hall stuck with the same basic design for his 1967 contender, the 2G.

THE MAIN DIFFERENCE BETWEEN the two cars was that the newer one would use Chevrolet's 7-litre engine. The 'big-block' unit had been available for some time, but was considered too heavy for use in racing. Now however Chevrolet's R&D department had built an alloy version which delivered some 575bhp. It was made available exclusively to Chaparral this year, though others would be able to get their hands on similar units a year or two later.

Hall decided to field only one car in the 1967 series, but the first part of the season was a disappointment. In spite of the light weight of the engine, and other efforts to pare pounds, the 2G was heavier than its predecessor, and even its greater power did not allow Hall to qualify higher than tenth in the series opener at Elkhart Lake. He hung in to take fourth place at the end, but then retired half-way through the Bridgehampton race, and gave the Mosport round a miss while he carried out more development work back at base. The result was second place at Laguna Seca, though only after several of the front-runners had retired, and the Chaparral was a lap behind the winning McLaren at the chequer.

**Chaparral's close association with General Motors meant their Chevrolet V8 racing engines, such as this one used in 1968, were better than anyone else could buy**

**The 2G devised for 1967 was really just a 1966-model 2E beefed up to take the power increase from the alloy 7-litre engine, available at this stage only to the Texan team**

But at Riverside the 2G finally came right. During the race Hall battled with McLaren for the lead, had a spell in front, and was only three seconds behind the New Zealander at the end. At Las Vegas, too, the car showed a competitive streak which was the equal of its predecessor's best 1966 performances. Here Hall qualified second-fastest to McLaren, and thus became the only man other than works McLaren drivers McLaren, Hulme or Gurney to have qualified for a top-three grid placing all series. The Chaparral again led the race, albeit after McLaren had problems, but the American car had troubles of its own, and did not last the distance.

### Cd GOINGS-ON

For the 1968 series Hall's fertile mind came up with a completely new approach. This time the aim was ultra-low drag – forget downforce, let's build a car that presents the smallest possible frontal area to the airstream. If it's as fast on the straights as it should be, it won't matter if it loses a bit of time on the corners.

By this time the wheels of a sportscar, which in earlier times caused the biggest disturbance to the airstream, were enclosed by full-width bodywork. But airflow over the top of the car was hindered first by the top of the driver's body, which jutted out into the airstream, and – at least in this era – by aerodynamic wings. Lose the wing and the car will be better streamlined. But what to do about the driver?

Hall's answer was to sit him beneath the top deck, with visibility provided by a clear nose-section in front of him and clear body panels alongside. A revolutionary approach, but it looked good on paper.

The dramatic Chaparral 2H was built on a narrow full-length monocoque chassis made from fibreglass (as with the very first rear-engined Chaparral). In another adventurous break with common practice the rear suspension was on the old-fashioned De Dion principle, but incorporated a 'load-levelling' system, a fore-runner of the 'active' suspension used by later designers. Power would come from an alloy-block 7.6-litre Chevy engine developed to produce a very healthy 680bhp.

The finished product was much lower and narrower than other Can-Am cars, but during pre-season testing at Rattlesnake Raceway serious axle-tramp was encountered, and the rear

Above **Hall should have been driving the revolutionary 2H in 1968, but prohibitive axle tramp meant the 2G saw another season**

Below **Revolutionary 2H car was designed to run with driver below level of top deck. But John Surtees refused to drive it before a more conventional seating position was devised**

Above **Surtees contemplates the 1969 Chaparral 2H. The body language suggests he wasn't entirely happy!**

Below **By the end of the 1969 season, further development had seen the 2H sprout bigger aerodynamic aids front and rear, but it was still not the success its predecessors had been**

Overleaf **Last of the line – the 2J 'sucker car', here with Vic Elford at Road Atlanta, where he outqualified the entire grid. The car was too revolutionary for officialdom, who banned it at season's end**

suspension had to be completely redesigned. The car would not race until 1969.

It was thus in the comparatively conventional 2G that Hall set off on the 1968 series. He held an initial third place at Elkhart Lake, but a couple of spins in the rain left him fifth at the end. At Edmonton he battled the works McLarens before brake trouble struck, costing him several laps. After rejoining, he reclaimed one of these, and set the race's fastest lap, but was still down in 11th when the flag came out. Then followed another clash with the orange cars at Bridgehampton, where the Chaparral sailed past first one and then the other to lead the race, before it was slowed by engine trouble which left it in second place at the finish. At Laguna Seca Hall became the first man all season to split the McLarens in qualifying, only for the starter to fail as the field lined up, and the Chaparral was wheeled away.

There was some consolation from third place at Riverside, and at Las Vegas Hall again qualified faster than everyone but the two works McLarens. However, he was delayed by the first-corner pile-up, and made several more stops later in the race. At last going well, though now miles behind, he came up behind Motschenbacher's private McLaren, and a misunderstanding led to a huge accident, in which both cars were destroyed. Hall was badly hurt, and a broken jaw and two broken legs meant it was many months before he was back at work.

His final placing in the championship was fourth, an improvement of one on his 1967 position.

## SURTEES GETS THE LOWDOWN

While Hall was recovering, he made the decision to retire from driving, but as soon as he could he resumed development work on the revolutionary low-drag Chaparral 2H. John Surtees was free to drive the new car, but used Hall's McLaren M12 in the

66

Above **Riverside, where Elford and the 2J were again fastest in qualifying. Speed was not matched by reliability however, and the auxiliary fan-motor failed on the very first lap**

Left **Twin fans in 2J's rear sucked air from underside of car in motion, forcing it down onto the road. It was this application of the laws of aerodynamics which caused offence**

first few races while the interrupted 2H test programme was completed.

The car had by now proved to be faster around Rattlesnake than any previous Chaparral, but then a new problem arose. Surtees simply refused to drive the car from its intended position under the deck. A hole had to be cut in the deck and the seat raised but, more seriously, it was back to the drawing board as far as the whole aerodynamic package was concerned.

The Chaparral 2H finally made its appearance in the fourth round, at Edmonton, still without a high wing, but with a low aerofoil on the back of the bodywork. After a delay caused by a broken throttle-spring, Surtees brought it home in fourth place. That was followed by fifth at Mid-Ohio, losing a battle for third with a couple of Lolas when the 1966 Can-Am champion had to stop for new goggles. Then at Elkhart Lake, after qualifying a lowly tenth, the 2H retired with damaged suspension. It was then returned to base for further development and missed two more rounds.

The 2H was back in action at Laguna Seca, now with a huge wing mounted just behind the cockpit, but little improvement was apparent, and Surtees could not match the pace of the fastest ten qualifiers. One imagines he was far from unhappy when it died on the pace-lap, and when it lasted only four laps at Riverside before its engine failed. The unloved 2H did not even turn up at the final round of the series.

The 2H was by far the least successful of all Chaparral designs. Hall and Surtees blamed each other for the failure, and they were probably both at least partly right.

### NEVER GIVE A SUCKER AN EVEN BREAK ...

The next – and last – Chaparral was just as revolutionary as its predecessors, but in a completely different direction again. The Fédération Internationale de l'Automobile, or FIA, the world governing body of motorsport, had now banned the use of suspension-mounted aerofoils in the Can-Am, and of any other 'movable aerodynamic device', so other methods had to be found to provide downforce. Most designers simply mounted their wings on the cars' bodywork, but Hall and his friends at General Motors took a completely different approach.

Chevrolet R&D had been carrying out their own experiments into aerodynamic principles, specifically the application of skirted fans as used on hovercraft and on an experimental French train. The first Chevrolet tests, using rudimentary skirts and with chain-saw motors to drive the fans, had proved sufficiently promising for a special 'vacuum traction vehicle' or 'ground effect vehicle' to be constructed for further research. As part of this programme they asked Hall to race the car.

The Chaparral 2J, which is what the car became, was a box-shaped structure built on a normal monocoque with conventional suspension and a fibreglass body, and using an 8.1-litre aluminium engine good for 700bhp. That was the standard bit: from there on it was all different.

Along the bottom of the car on both sides and across the front and rear were tough plastic curtains linked direct to the suspension and designed to maintain a minimal gap to the ground. The other significant component was an air-cooled 275cc Rockwell snowmobile engine with belts which drove two vertical fans in the stern.

The idea was that the fans would draw air from the virtually sealed underside of the car, and 'suck' it onto the road. Cornering would thus be improved without the need for a high wing.

Of course, as soon as Chaparral's horrified competitors saw the car on the track – especially after noting its lap-times – they complained that if the fans weren't movable aerodynamic devices, the skirts were. Either way, they said, the car was illegal under the rules. Hall, who had had the project approved by SCCA officials before the season started, preferred to talk about 'static pressure devices', and the car was allowed to run.

The Chaparral 'sucker car' was not in the field for the two Canadian rounds which opened this season's schedule, but was wheeled out before an incredulous audience at Watkins Glen. And with reigning world champion Jackie Stewart at the whe el it was competitive, too, for in spite of being by far the heaviest car in the field, it lapped faster in qualifying than everything except the works McLarens. In the race Stewart held third place until his pace was hampered by brake troubles, and the 2J was eventually retired.

Further development was needed if the Chaparral was to become a McLaren-beater, so it was back to base for more work. Much of the development driving was carried out by Englishman Vic Elford, who had been a mainstay of the Porsche sportscar team and in 1968 and 1969 a Grand Prix driver, and who had raced various other cars in the 1970 Can-Am series.

The car's next race appearance was in the seventh round of the series, at Road Atlanta in Georgia. And it was with Elford at the wheel, for Stewart's arrangement had been for one race only. Elford's efforts and those of the rest of the team were duly rewarded with fastest time in qualifying. And that was a considerable achievement, because you had to go back to Riverside in 1967 to find a car other than a works McLaren starting a Can-Am race from pole position. Elford made a slow start in the race however, and could not close the gap to the McLarens. He did eventually get to second place, but soon afterwards the 2J was in the pits for attention to its fan-motor, rejoining to finish sixth.

After missing another round Elford qualified on pole again at Laguna Seca, but the effort had destroyed the engine, and the car was unable to start the race. Not only was he then fastest in practice for the Riverside finale, but he did it by two seconds from Hulme. Halfway around the first lap however the fan-motor went again, and that was that.

There is no doubt that the Chaparral 2J had, by the end of the season, established itself as the fastest car in the championship, and might have been expected to dominate in 1971. However, the FIA had been viewing the design with some alarm, and at the end of the season overturned the SCCA's ruling that the 2J's skirts were not movable aerodynamic devices. The sucker-car could race no more.

Although not perhaps obvious at the time, this decision marked the beginning of the end of the Can-Am series. It contradicted the intention of encouraging innovation by formulating a minimum of rules, and led the way to more restrictions being introduced in later years. It was all too much for Hall, who closed down his Chaparral operation and announced his withdrawal from the series. Can-Am had lost a marvellous competitor, and endangered an important principle.

CHAPTER FIVE **CAN-AM**

# Lola and other BRITISH CHALLENGERS

In terms of numbers, if not success, Lola remained the only serious British-based competitor apart from McLaren throughout the early years of the Can-Am. Between 1966 and 1970 the two major British suppliers were responsible for approximately equal numbers of cars on the average Can-Am grid, and between them usually filled 75% of the grid places, sometimes as much as 85%.

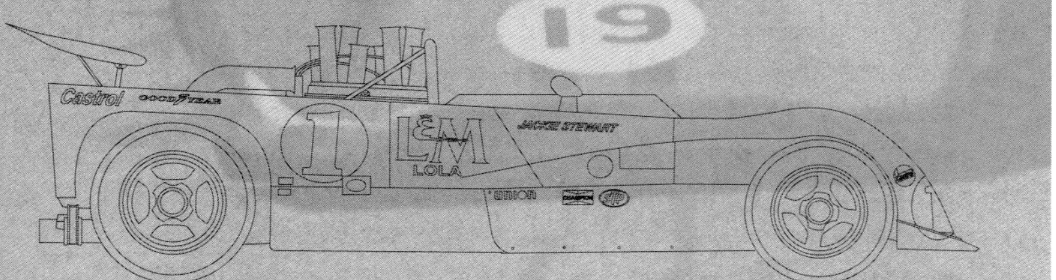


Many other British constructors looked enviously at the Can-Am market but only two, BRM and March, went so far as to build cars for the series. Both operations had their hands full with other formulas however, and neither persevered.

In the early races the occasional Lotus, Brabham or Merlyn had made an unremarked appearance, but all had been built for earlier racing, and were no longer competitive. Not British, but then not American either, was the Matich-Repco, a nicely presented contender from Australia which ran in most of the 1967 races. Its 4-litre Repco engine lacked power however, and the venture made no impact on the series.

Lola would stick with the series until 1972, but in spite of flashes of competitiveness. not even they were able repeat the success they had enjoyed in 1966.

For 1967 they produced a revised Mk3 version of the T70, and found plenty of good people who wanted to run it. John Surtees, Dan Gurney and Mark Donohue, the three top Lola men of 1966, each had one while Roger Penske, as well as fielding a car for 1967 USRRC winner Donohue, also entered a back-up car for George Follmer.

Surtees again drove a works development Lola this year but even his undoubted sorting skills and driving talent were not enough to match the pace of the McLaren M6As. The first two rounds brought only a third and a fourth, and the next three resulted in retirements. Only in the final round at Las Vegas did everything come together for Surtees, and then only after he had ditched the Mk3 and dusted off his 1966 Mk2. Even so, both works McLarens had to strike trouble in the race before Surtees was able to give Lola its only win of the year. Although well behind on points, he salvaged third behind McLaren and Hulme in the final championship standings.

Gurney again used Weslake-developed Ford engines in his car, this time a 6.2-litre version of the new 5.8-litre production

**Lola T160 was unsuccessful replacement for successful T70. Chuck Parsons in the car entered by Lola importer Carl Haas did not finish the race at Bridgehampton in 1968**

Lola suspension was more or less conventional for the day, but some later models than this T70 experimented with horizontal front springs

unit. He actually proved faster than Surtees over the full series, and in fact qualified for the first three races faster than everyone bar the invincible M6As. He then split them at the next meeting, and actually qualified on pole at Riverside before dropping back to third again at the final round. This form was continued in the races, in most of which Gurney showed himself to be on equal terms with anyone. He even led the McLarens on occasion, But form is one thing and reliability another, and poor Gurney failed to finish a single race.

Parnelli Jones also campaigned a Lola-Ford again, this time with a development of the DOHC Indy engine. He did not join the series till its halfway point, but qualified well up in all three races that remained. At Laguna Seca he challenged for the lead, and in both the next races actually succeeded in getting to the front. But the record-book shows only a single fourth place and two early retirements.

Donohue used a big-block Chevrolet engine in some races, and a smaller unit in others, but was unable to match his 1966 reliability record. In the races he finished however he took two second places and a third, and was leading the Las Vegas round on the last lap when he had the wretched luck to run out of fuel. Nevertheless, his results left him with the same season points total as Surtees. His Penske team-mate Follmer, with two

Below **Lola T160 was the unsuccessful replacement for the T70. Chuck Parsons in the car entered by Lola importer Carl Haas did not finish the race at Bridgehampton in 1968**

Overleaf **In 1969 Haas and Parsons campaigned the T163, a much stronger contender than the T160, but still not a threat to the McLarens. Nevertheless, Parsons finsihed third in the championship**

10
SIMONIZ
4-1-1 special

10
SIMONIZ
4-1 special

thirds and two sixths, finished sixth in the championship, while other Lola drivers in the points included Peter Revson and Lothar Motschenbacher – both of whom had defected from the McLaren ranks.

### SOMETIMES, THE OLD ONES ARE THE BEST ...

Lola had been beaten fair and square by McLaren in 1967, so for the following year came up with a new model, the T160. This had a stiffer full-length aluminium monocoque with magnesium bulkheads but, although lighter than the T70 on whose design it was based, it was still heavier than the McLaren opposition.

Team Surtees, having now broken its links with Broadley, ran its own programme this year, with a highly modified T160 running a 7-litre aluminium Chevrolet engine with special Weslake heads. The car was also fitted with a Chaparral-style high wing but, in spite of the effort that had gone into its development, was pretty much a failure. Surtees retired in the first two rounds and, even after heavy revisions, struggled to qualify for the third, in which he lasted only two laps. He then missed a couple of rounds to develop the car further, but could still only qualify eighth at Riverside and, after retiring once again in that race, did not bother with the final round.

Dan Gurney, now campaigning McLarens on a regular basis, also ran a Lola T160 in a few races for his protegé Swede Savage. Fitted with a 5.3-litre Ford engine it picked up a fourth place before Gurney took it over for himself. He then installed a 7-litre aluminium Ford engine built up by Holman and Moody, and good for around 580bhp. At Riverside Gurney was competitive while the car lasted, and at Las Vegas showed even better. He was pressing Hulme's McLaren for the lead here when a half-shaft broke on lap 15.

In that same race Mario Andretti drove a similar car, and ran second from the start, but was forced to make a long pit-stop which left him 12th at the end. Chuck Parsons in a Chevy-powered version entered by US Lola importer Carl Haas managed a couple of top-six placings during the season, and so did privateer Sam Posey with a similar car.

Significantly perhaps, Andretti in a theoretically obsolete T70 powered by a 5-litre DOHC Ford engine had proved in earlier rounds to be just as fast as the newer cars. He challenged McLaren on a wet track at Elkhart Lake before falling back, and was in fact still in third place when the car expired two laps from the end. Follmer in another T70-Ford showed strongly in the Laguna Seca downpour, before crashing out near the end of the race. But he kept going to take a fine second place at Las Vegas.

**The T220 model of 1970 was an improvement on earlier cars, but still not good enough to match the McLarens. This is Peter Revson in the Haas entry at Watkins Glen**

LM
LOLA
26

**Second-hand Lolas provided less wealthy teams with the chance to take part, if not to gain much in the way of placings. Bob Bondurant in a two-year-old T160 at Road Atlanta in 1970**

### DRIVERS CHANGEOVER

Lola was even less of a presence in the next year, for the failure of the 1968 car meant that none of the big names was seen behind the wheel of its 1969 replacement. Parsons and Revson, the principal Lola regulars this year, were fine drivers, but not in the same class as Surtees, Gurney or Andretti.

Nevertheless, the T162 and T163 updates of the T160 proved surprisingly competitive, and when Roger Penske joined the series at Mid-Ohio, his man Donohue qualified third fastest to the works McLarens. Donohue was holding the same position in the race when a half-shaft broke. After that one-off showing however the Penske team withdrew from the series to concentrate on its busy USAC and Trans-Am campaigns.

Parsons contested the full series in Haas-entered cars, powered this year by 7-litre Chaparral-built Chevrolet engines. He qualified in the top six places on several occasions and his race results were good, too, with seven finishes in the top half-dozen. These included three third places behind the unbeatable McLarens, and on another occasion he was battling with Donohue for another third when he retired. All this left him third on points for the series, behind McLaren and Hulme. Nothing more could reasonably have been asked.

Revson took a while to find his pace, but qualified fourth-fastest for a couple of the end-of season races. At both Michigan and Laguna Seca he held third place behind the McLaren M8Bs, but suffered more than his share of unreliability, and at the end of the season had only a fourth and a fifth to show for his efforts.

### THE NEARLY CAR

The ban on suspension-mounted wings applying for 1970 led to an all-new Lola, the T220. Apart from an unusually short wheelbase, this was a fairly conventional design, built on a full-length monocoque chassis. Competition would be as tough as ever however, not only from the perennial pace-setters McLaren, but also from the Chaparral 'sucker-car'.

Revson drove the Haas car this year, and qualified well for all the opening rounds in the face of such opposition, only to suffer a number of retirements. But he placed the dumpy car an excellent second at Mid-Ohio, and led the entire field at Elkhart Lake for a couple of laps. He was again in front at Road Atlanta when he crashed off the road.

It is presumably no coincidence that the car built to replace the original T220 was ten inches longer. First time out, on the Donnybrooke circuit at Brainerd, Minnesota, Revson planted it squarely on pole, only to have the throttle-cable break on the opening lap. A stop for repairs dropped him to 17th, but he battled his way back to third at the end. He held third place again in both the closing rounds, staying there till the end at Laguna Seca but retiring at Las Vegas. His tally of one second

place and two thirds was a poor reflection of the car's potential, as was his eighth place in the championship standings.

Well ahead of him on the points table, in fourth slot, was veteran Dave Causey. Driving Revson's T163 from the year before, Causey took a second, a third, a fifth and several lesser places over the season, while Parsons in a similar car took two top-six placings and Gary Wilson, in yet another, three. Former works Ferrari driver and later racing-school proprietor Bob Bondurant qualified an even older T160 in the top half-dozen a couple of times, and finished a race in second place on one occasion, but otherwise suffered several retirements.

Top **The T260 was Lola's first really competitive contender since the T70. With world champion Jackie Stewart as driver, it was as good as the McLarens, as here at Watkins Glen**

Above **Stewart upset the 'invincible' McLarens by taking pole position at Watkins Glen in 1971, but the car did not last the race. Note unique nose treatment, and cutaway behind front wheel**

### HIRING THE BEST

The following year brought Lola its most successful season since 1966. The T220 was put to one side, and Eric Broadley and his team – which this year included Bob Marston and a youthful John Barnard – came up with the T260.

Another full monocoque, it featured side radiators, inboard brakes and horizontal springs. The body was designed with the emphasis on low drag rather than high downforce: in this it followed not only the Chaparral 2H, but also Lola's own successful 2-litre sportscars and those of their class rivals Chevron. Downforce was provided at the front by a series of holes in the bodywork, with the rear aerofoil mounted forward of the rear wheels. During the season however this was moved back, and a large wing hung onto the front of the car. Power came from an 8.1-litre Chevy.

Convinced he at last had a serious challenger to the McLarens, Broadley's US agent Carl Haas took a deep breath and hired Jackie Stewart, on his way to his second world championship, to do the driving.

And it almost worked. As far as lap-times were concerned, the combination of Stewart and the T260 was the equal of the works McLaren M8Fs. Stewart showed this right from the start

Left **The small rear aerofoil on the T260 was initially mounted further forward (as pictured) than on rival cars, but was later enlarged and repositioned behind the rear wheels**

Below **The nose of the T260 was designed to provide low drag rather than high downforce. By season end however, a big wing was extended out of the front of the car**

Right **Rear suspension of the T260 was conventional, though top radius arms ran all the way to the bulkhead. Coil-spring damper units were mounted at the usual angle, unlike the horizontally-mounted fronts**

Right **Moment of glory for the Lola T260 as Stewart follows the Watkins Glen pace-car from pole position, McLarens of Peter Revson and Denny Hulme alongside and behind**

of the season when he was fastest qualifier at Mosport. Another pole came at Watkins Glen, and seconds and thirds at most of the other meetings.

In racing conditions, too, the T260 was a match for the McLarens, and Stewart headed them for at least part of the distance in several rounds. There were the inevitable retirements, but in the second round of the series, at St Jovite, Stewart gave Lola its first Can-Am victory since 1967, and repeated the effort at Mid-Ohio. Backed by seconds at Edmonton and Laguna Seca, the Scot finished third on the championship table, with only the works McLaren drivers ahead.

The 1971 customer Lola was the T222, a development of the previous year's T220. In one of these Japanese driver Hiroshi Kazato achieved five placings in the top ten over the season, though only two were in the first six.

## FINAL EFFORT

Encouraged by coming so close to real success in 1971, Lola came up with another all-new car for the following year. The T310 was longer – in fact the longest of any Can-Am car – and wider, but also lower, the engine being tilted over slightly. The shovel-nosed body was developed in the wind-tunnel, and was equipped with a low aerofoil mounted well back. But little time was available for pre-season testing, and the T310 was always at least one step behind the game.

Driven by David Hobbs, it proved able to qualify in the top five or six, but its best placings were a fourth, two fifths and a sixth. A challenge to the Porsches and McLarens it was not.

Charlie Kemp was the top privateer of the season, with three top-six finishes in a T222.

Lola then withdrew from the series, leaving it to independent drivers in ageing cars to fly the flag. Bob Nagel managed no fewer than five top-six placings in a T260 in 1973. These included a third, but his results were as much a reflection of the falling standard of the series as anything, as indeed were the four fourth places he took with the same car in 1974. John Gunn in another T260 was third in one 1974 race and fifth in another. Nagel's final placing in the 1973 title contest was fifth, and he was fourth in 1974, with Gunn one place further back.

## IN AND OUT

BRM and March had both appeared with new Chevrolet-powered cars in the 1970 series, shone briefly, then disappeared.

BRM, until now involved exclusively in Formula 1, used an aluminium monocoque designed by Tony Southgate. Although departing little from standard layout, the car's body-shape was the result of extensive wind-tunnel testing. The company's own

development of the aluminium-block Chevrolet engine was used, first at 7.0 litres and later 7.6, coupled to a Hewland transaxle.

Driven by Canadian George Eaton, the BRM P154 retired in two of the first three rounds it contested but was third in the other, and at Road Atlanta led the entire field in the middle stages before going out with low oil-pressure. Eaton retired in his two other races while Pedro Rodriguez, who drove a second P154 in three races, took a third, a fifth and a ninth.

The car reappeared in revised (P167) form in 1971, but did not join the series until the last two races of the year. By then it had won two races in Europe, but the competition was tougher in America. Endurance racing star and sometime F1 driver Brian Redman brought it home in fourth place at Laguna Seca while BRM F1 team member Howden Ganley, who took it over for the Riverside race, went one better, and finished third.

But by then the company had tired of sportscar racing and, saddened by the loss of their two star drivers, Siffert and Rodriguez, turned all their attention back to Formula One.

### FAIRLY QUICK MARCH

March, in their first season, were involved in an ambitious production and racing programme which included F1, F2 and F3. But part of their contract with Chris Amon required them to build him a Can-Am car, and the 707 was the result. Its design was the responsibility of former McLaren man Robin Herd, and featured a simple aluminium semi-monocoque using the engine as a stressed member but with magnesium bulkheads. Suspension was also conventional, while power came from a Chaparral-built aluminium-block 685bhp Chevy driving through a Hewland LG600 gearbox.

Outwardly, it differed from other Can-Am cars in employing F1-style adjustable front wings.

With so many other things going on at March in 1970, it is perhaps not surprising that the car did not join the series until the eighth round, at Donnybrooke. But the 707 was on the pace from the start, Amon qualifying third, and getting between the works McLarens for a short while during the race. He was still in third place three laps from the end when the car stopped with a broken fuel-line, which meant he was classified fifth. This was followed by fourths in the car's other two races, at Laguna Seca and, after losing a place with a last-minute refuelling stop, at Riverside.

The first March suffered a weight disadvantage over its rivals, but no-one was too concerned, as it was planned from the start as a preliminary to an all-out assault on the 1971 series. However, although an updated version, the 717, duly appeared, it did not race in the Can-Am. The 707 was seen, but in private hands was very much an also-ran.

Right **BRM P154, here with Canadian George Eaton trying hard at Watkins Glen in 1970, might have done better if factory had not been concentrating so much on Formula One**

98

Top **March ran the 1970 series as a tryout for a stronger campaign in 1971. The original car was built as 717 for a customer, but raced only in Europe. It still races in Europe: these pictures show it at Spa in 1999**

Above **By 1971 the big-capacity alloy-block Chevrolet V8 engine was virtually unchallenged in Can-Am**

Right **With the March factory deciding not to continue its Can-Am programme, the 717 displayed no radical departures from the 707, and did not gain much success even in the races it ran in Europe. An imposing lump nevertheless. In 1971 (hence the 717 designation) the car was modified and the nose shortened to become a group 7 car, driven by Helmut Kelleners**

STELLA
ARTOIS
FINA
SOIR
41
42

# McLaren without McLAREN

No matter what anyone else tried – Lola, March BRM, even Chaparral – the pale orange works McLarens just kept on winning everything in sight. One or other of the challengers might on occasion have looked like denting the team's clockwork efficiency, but each year the New Zealand-run British-based company bounced back with a better car and maintained their supremacy. Having been all but unbeatable in the 1967, 1968 and 1969 series, they seemed certain to continue their domination in 1970.

**Even without Bruce McLaren, the cars that bore his name continued to dominate the Can-Am. Hulme, here winning the Laguna Seca round, was 1970 champion in M8D**

But on 2 June 1970, less than two weeks before the opening round of the new series, Bruce McLaren was killed testing his latest M8D Can-Am car at Goodwood. His close-knit team was shattered, but determined to proceed. To add to the disarray the team found itself in, Hulme had burned his hands in an accident at Indianapolis three weeks before. But he insisted on leading the McLaren Can-Am defence, and brought Dan Gurney onto the side to drive the other car.

The M8D was evolved directly from earlier M8 variants, its most noticeable structural difference being that it was wider. This was the year that suspension-mounted aerofoils were banned, and the M8D used high tail-fins with a low wing attached between.

To counter the anticipated loss of lap-times brought about by the new regs, the team concentrated on finding more power. For the first time this year McLarens had access to Reynolds silicon-aluminium Chevrolet blocks, which were carefully prepared by George Bolthoff and his team, now based in America again. In 8-litre form these engines produced some 700bhp, but for reasons of reliability 7.6-litre versions were used for the campaign. The team also reverted to 7-litre units on some occasions while overheating problems in the bigger engines were sorted out.

Before the Can-Am series the wings on the race cars were braced to the chassis, as it seemed McLaren's accident had been caused by the airstream lifting the wing at speed, and tearing the rear bodywork off.

The season started in its familiar pattern, as the works McLarens set the two fastest times in qualifying for the first three rounds. Gurney earned himself pole starts in two of these races and Hulme in the other. The American won the Mosport and St Jovite rounds, with Hulme, still in considerable pain with his burned hands, third in the first but retiring in the second. The New Zealander then won the Watkins Glen round, in which it was Gurney's turn to strike trouble. But Hulme and the rest of the team had shown that, even without their leader, the McLarens were just as effective as they had ever been. One can, however, imagine the mixed feelings with which fans and insiders alike watched the orange steamroller continue on its way without Bruce McLaren.

But then before the next race Gurney's oil sponsors, Castrol, made it clear they did not like him racing (and winning) in Gulf-backed McLarens, and suggested he might like to cease. Two weeks after the Glen therefore, it was Peter Gethin who drove the second works McLaren.

Although he had shone in smaller cars and had won the 1969 European F5000 championship (in a works McLaren), Gethin was not in the same class as Gurney or McLaren, which placed even more responsibility on Hulme's shoulders. The 1968 Can-Am champ rose to the occasion by taking four more pole positions over the remainder of the series and was second-fastest qualifier for three other races. These efforts translated into wins at Edmonton, Mid-Ohio, Donnybrooke, Laguna Seca and Riverside, and his second Can-Am title.

Gethin provided useful backup, qualifying in the top half-dozen for all of his races. He never managed a pole start however, and was second fastest at only two rounds. His race showings, too, were disappointing: he won the Elkhart Lake round, but only after Hulme, who had crossed the line first, was penalised for having received a push-start after spinning to avoid a back-marker. Otherwise the Englishman's only placings were a couple of seconds, which left him third on points.

There were of course, as always, numerous private McLarens in the series. They were led this year by Lothar Motschenbacher, who started with one of the 1969 factory M8Bs but later drove an M12 and later still an M8C, which was the 1970 customer model, a rebodied version of the M8B. He qualified consistently well – twice in the top three – and finished in the top six places in six races, including a second and three thirds. It was he who split the works drivers in the series points standings, a fine

**Peter Gethin was drafted into the team at short notice and finished third in the championship, winning one round. The 'fences' on the M8D front wings are clearly visible**

Above **Roger McCaig in a private McLaren M8C has the works M8Ds of Hulme and Gurney bearing down on him at St Jovite. Gurney soon had to leave the team for contractual reasons**

Right **At Laguna Seca in 1970 Gethin equalled Hulme's time in qualifying, and was following his team-mate until he spun off on oil. The car wasn't damaged, but could not restart**

showing and fitting reward for one of the Can-Am's most loyal supporters.

Bobby Brown drove the ex-Gurney McLaren-Ford M6B this year and four times finished in the first six, but in his best showing crashed into the wreckage of another car when holding third.

## REVSON TRIUMPHS

For the 1971 McLaren M8F, Gordon Coppuck devised a chassis which was stronger, being made of heavier-gauge aluminium, and three inches longer than its predecessor, and also moved the rear brakes inboard. Rear rim-widths were now up to seventeen inches. The tops of the body-sides were fitted with aerodynamic fences which rose into fins which, as with the M8D, also served as mounts for the rear wing.

Gary Knutson had now returned to take charge of engine development at the US facility, and looked at two new engine possibilities. One was based on Chevrolet's own 8.1-litre alloy block, which had iron liners and gave 740bhp, while the other was Reynolds Aluminium's linerless (and therefore lighter) version of 7.8 litres, which was also run at 8.3 litres, and produced a similar figure.

More and more new challengers were appearing with every year, but the factory McLarens were again the pre-series favourites. Hulme's team-mate this year was American Peter Revson, an experienced Can-Am competitor who had joined the team for USAC racing, and was also contesting his second F1 season, with Tyrrell (his first had been as a privateer back in 1964). The new team member proved a valuable recruit, and was immediately able to return lap-times much closer to Hulme's than Gethin's had been the previous year.

Revson actually outqualified Hulme once in the first half-dozen rounds, then proved how well he was adapting to his new

team by doing it three times in a row, before the senior team member turned the tables again for the last of the ten rounds. In spite of strong opposition from Stewart in the latest Lola, Hulme ended the season with five poles under his belt and his team-mate with three. On seven of those occasions the McLaren duo set the two fastest qualifying times.

Qualifying times are an indicator of potential, but it is results that count, and once again the works McLarens record was unmatched.

Hulme and Revson finished first and second in the series-opening Mosport race, but Stewart held them to second and third at St Jovite. Revson then led Hulme over the line to win both Road Atlanta and Watkins Glen, but both drivers struck trouble at Mid-Ohio, the New Zealander retiring and the American placing seventh. That was after a universal-joint failed when he was in the lead just nine laps from home.

But the M8Fs then won every remaining round. Revson was first at Elkhart Lake, where the other car retired. At Donnybrooke it was Revson first, Hulme second, but at Edmonton it was the American's turn to retire, leaving Hulme first. Revson bounced back to win at Laguna Seca, though in controversial circumstances. He was black-flagged on the last lap for dropping oil, but continued anyway, and led Stewart's Lola over the line. The Scot's entrant protested, but the stewards confirmed the McLaren driver as the winner – though they did fine him.

Hulme was third there, but the season ended with another 1-2 at Riverside, Hulme winning from Revson. Once again,

Overleaf **American Peter Revson had his reward for years of service in uncompetitive Can-Am cars when he drove a work McLaren M8F in 1971, and beat Hulme to win the title**

REYNOLDS
ALUMINUM
7
GOOD YEAR
McLAREN CARS
CHAMPION
GOODYEAR
Gulf
REYNOLDS
ALUMINUM

Gulf

**Hulme waits for the fuel to go into his M8F at Watkins Glen in 1971. This was the year the McLaren domination was first challenged, and the orange cars were lucky to take the title**

there were works McLaren drivers in the first two places on the points table at season end, Revson on top and Hulme second.

Motschenbacher, the only man to have contested very single Can-Am race since 1966, this year campaigned an ex-works M8D, and was again the best of the McLaren privateers. He regularly qualified with a time among the top half-dozen, and wound up fifth in the championship standings, from two third places, a fourth, a fifth and a sixth. He also entered a modified customer M8E in some races for Bob Bondurant, who was not much slower, but could produce only a single fourth place for his efforts.

Oscar Koveleski campaigned another ex-works M8D, and then put former North American F5000 champion Tony Adamowicz in it and was rewarded with a third and a fourth. Vic Elford in an M8E entered by Roy Woods also took a third and a fourth, and Sam Posey, standing in for the English driver in a later race, brought the team another fourth. Among other McLaren privateers Gregg Young took a third while Chuck Parsons and Swedish-born Canadian John Cordts, Can-Am veterans both, managed a pair of top-six finishes apiece.

## TURBO, OR NOT TURBO?

Since Bruce McLaren's death the team that bore his name had won two more Can-Am titles, but there would be no more. In 1972, as will be detailed in a later chapter, the turbocharged Porsches swept all before them and the new McLaren M20 was reduced to the ranks of the also-rans.

The team had in fact experimented with their own forced-induction Chevrolet units, and John Cordts had tried a similar but privately-developed set-up at one of the 1971 rounds, but abandoned it after practice.

It was with the use of turbocharged engines in mind that Gordon Coppuck designed the 1972 car. The M20 was another two inches longer than its predecessor and had bigger brakes. Side-mounted radiators were used, which enabled an aerofoil to be mounted between the front wings, with a full-width rear wing.

The team also had another ace up its sleeve. Revson had decided he wanted to concentrate on Formula One, and Jackie Stewart was signed to drive alongside Hulme, which would have given McLaren by far the strongest driver line-up in the series. As it happened though, the Scotsman missed the pre-season test programme through ill health, and then decided not to do the series. So Revson returned after all. By this time the team had decided not to proceed with the turbocharged engine programme, and adapted the M20 to take the more familiar normally-aspirated Chevrolet engine instead. It was not enough.

Above **McLaren built the M20 to run turbocharged engines in 1972, but abandoned the concept before the season started. They lost the championship to Porsche (who used turbo engines)**

Right **M20 was the last McLaren Can-Am design, but won only two races. Hulme, here at Mid-Ohio, was the successful driver, but was beaten in the championship race**

The defence didn't start too badly. Hulme won the opening round, at Mosport, with Revson classified third, even though his crankshaft broke two laps from the end. After both retired at Road Atlanta, the McLarens bounced back to take the first two places at Watkins Glen, Hulme again being the winner. Gregg Young's pair of ex-works M8Fs also showed well in these early races, the team owner himself taking a second and F1 driver François Cevert in the other car a third.

But then the Porsche Turbo bandwagon began to roll, and although Hulme managed one more pole start, his best results over the remainder of the season were a second and a fourth, while Revson added only a second and a sixth. Both drivers suffered a string of retirements as the M20s surpassed their stress limits in an effort to match the pace of the German cars. In fact Cevert won the Donnybrooke round, and took a second and a third as well, and Young added a fifth to his earlier score (before writing the car off in a later race).

In desperation the works team bolted a 9.3-litre engine into one car and, with 785bhp at hand, Hulme split the turbo Porsches in qualifying for the Riverside round. But the engine

**David Hobbs in an ex-works M20 in 1973. This was often the fastest normally aspirated car, and brought Hobbs enough places for seventh in the series standings**

was not expected to hold together long, and it didn't. Hulme was a poor second in the title race, with Revson down in sixth, one place behind Cevert. The previous year's M8F had meanwhile been adapted for production as the M8FP, and several examples were seen during the season. Gary Wilson, Cordts and Motschenbacher all managed places in the top six in older cars, the latter on three separate occasions.

## PRIVATEERS PARTY ON

That 1972 defeat was enough for the new people who were now controlling McLaren fortunes, and the team announced that it would not be supporting the 1973 series. Porsche's budget in 1972 had reputedly been five times what McLaren had been used to spending, and the Germans also had at their disposal an entire factory and high-tech engineering facility to develop the new technology. McLaren simply could not compete with that. Nor, perhaps, could any other sportscar manufacturer with the possible exception of Ferrari: and their interests lay elsewhere.

Some of McLaren's customers had more confidence, however misplaced. Mario Andretti had the 7.6-litre Reynolds-Chevy engine in one of the previous year's M20s adapted to run a twin-turbo set-up which apparently gave over 1000bhp. This was harnessed to a Weismann transaxle, but the car missed the first 1973 round and could only qualify sixth for the next, at Road Atlanta. Andretti was unhappy with the car's overall preparation, and never raced it. John Cannon did get it to the grid, but Andretti's judgment had been right. The car was not a success.

Roy Woods ran another ex-works M20, though with normally-aspirated engine, over the entire season, and this proved a more viable contender. Regular driver David Hobbs qualified it in the top half-dozen places for several races, and finished second at Watkins Glen, and fourth in a couple of other races. Derek Bell in an M8F was fourth in one round, while Scooter Patrick, Bobby Brown and 1970 North American F5000 champion Cannon were all regular place-getters with similar cars. By 1974 the Can-Am Challenge no longer enjoyed much credibility on the international racing scene, and with Porsche joining McLaren on the sidelines it was left to the local Shadow team to dominate the abbreviated series. Faithful McLaren drivers Patrick and Cordts as often as not qualified third and fourth behind the American cars, but had more than their share of unreliability. Patrick did however pick up a couple of third places, and won the Elkhart Lake round which brought the series to a close, with Cordts second in the first McLaren 1-2 for more than two years. Patrick was in fact third on points behind the Shadow drivers. Motschenbacher also managed a couple of top-six placings with another M8F during the year, and was sixth in the championship.

Castrol

CHAPTER SEVEN **CAN-AM**

# Continental CHALLENGERS

Throughout the Can-Am era, the two big European names in the world sportscar championship were Ferrari and Porsche: as they had been since the late 1950s. Both also turned their eyes on the North American series, and at different times built cars for it. Ferrari – like BRM, March and most Lola models – flattered only to deceive, but Porsche persevered and ultimately achieved a McLaren-like domination of the class.

**Ferrari 612 made its first appearance of 1969 here at Watkins Glen, where Chris Amon lapped faster in qualifying than all but the works McLarens of McLaren and Hulme**

FERRARI'S US AGENT for many years had been Luigi Chinetti, whose North American Racing Team was well-known as an entrant of the Italian cars at Le Mans and elsewhere. In 1967 NART prepared a rebodied 4-litre 330 P3/4 model and ran it for works driver Ludovico Scarfiotti in the second and third Can-Am races of the year. It finished seventh at Bridgehampton but retired at Mosport.

The parent company's interest in the Can-Am series had received considerable impetus from the arrival at Maranello that year of Chris Amon, previously an enthusiastic member of the McLaren effort. Independent of NART's programme, the factory had gone to work on a pair of Can-Am cars of their own. These were lightened and rebodied 330 P4s with special 4.2-litre V12 engines giving around 480bhp.

They appeared first at Laguna Seca, but apart from introducing an excitingly different exhaust note as counterpoint to the American V8s, did not make much of an impression on the established teams. Amon in the faster of the two red cars qualified 16th, with his team-mate Jonathan Williams four places further down the grid. The Ferraris were equally outclassed in the race, but Amon kept going as others fell out and was in fifth place at the flag. Williams was eighth.

Amon went on to finish eighth at Riverside, and was in fifth place four laps from the end of the Las Vegas round when he retired. The other car did not finish either race.

## MADE-TO-MEASURE, BUT TOO HEAVY

Clearly something better was needed. Back in Italy, work began on a purpose-built Ferrari Can-Am car. Designed from scratch on a multi-tubular space-frame, the dramatic 1968 car had a specially-built 6.2-litre V12 engine with four-valve cylinder-heads which, running on fuel-injection, apparently produced exactly 100bhp/litre. A very wide aluminium body was built

Above **Following Surtees in the Chaparral team's McLaren M12 at Watkins Glen must have set Amon thinking about the handling benefits that a high aerofoil would bring**

Right **Ferrari's 1969 challenger was notable for its uncluttered lines, but confined cockpit caused driver discomfort from high temperatures, and the car did not handle well**

with a cockpit-adjustable wing (still allowed at that time) mounted on the chassis behind the driver's seat. Unfortunately, Ferrari's commitments in world championship Formula One and sportscar racing meant work on the 612P was delayed, and the car did not appear until the very last round of the 1968 series. In the meantime NART fielded an open-bodied 365 for Pedro Rodriguez in the Elkhart Lake round. But he could finish no higher than 13th, and the car was not raced again.

The 612's debut at Las Vegas was, after the long wait, another disappointment. In spite of its healthy power output, the car was still heavier than the McLarens, and that factor, coupled with inadequate braking and poor handling, meant that Amon was unable to qualify higher than ninth. Whether it had been improved between qualifying and the race is unkown as the car was eliminated by an accident on the opening lap.

## AMON'S BABY

For the following year the car was rebuilt, and in the process shed a considerable amount of weight. Power was up around 650bhp, and suspension and bodywork were revised. It was funded by Amon, who unfortunately received little factory help in its development. By and large however, the car was the fastest thing in the series – apart, of course, from the McLarens. Amon frequently qualified it third-fastest, though it handled properly only after a hub-mounted wing was added. The car also suffered from problems with its oil system, but managed a second place finish and two thirds before unreliability set in.

For the Riverside round Ferrari produced a 6.9-litre engine, but Amon was still unable to qualify higher than third, holding the same place in the race before striking trouble. The big engine blew in practice for the Texas finale and the 6.2 was put

16

16
16

Above left **Amon kept the Ferrari up with the McLarens for much of the Watkins Glen race, but finally had to ease up to save the big V12 engine as it started to overheat**

Below left **By Michigan Amon had fitted the Ferrari with an aerofoil of its own, and was at last happy with the handling. Car did not start this race after the engine seized in practice**

back in for the race. Once again Amon ran behind the fastest McLarens. Once again he retired. His sixth placing on series points was hardly a true reflection of the car's potential.

A 3-litre endurance-racing Ferrari 312P had appeared at the Bridgehampton round, Rodriguez finishing in fifth place.

Ferrari fortunes, in the main, continued to rest with the drivers of endurance-racing cars for the next couple of years. In 1970 Mario Andretti gave a 512S a run at Watkins Glen, where the race shared the bill with a round in the world endurance championship. He qualified the 5-litre V12 to start fifth on the grid, and finished in the same place in the race. It has to be mentioned, however, that there were three endurance-racing Porsches ahead of the red car. Rodriguez contested a couple of later rounds with a similar car, as did American amateur Jim Adams, though neither achieved much in the way of placings.

For the last three rounds of the series Adams wheeled out the car Amon had run in 1969, now fitted with a 512 engine for reliability. His only placing was a fourth, followed by an eighth the following year. Sam Posey, a very experienced Can-Am competitor and by now one of the leading names in American F5000 racing, took a sixth with another 512 in 1971, and Swiss driver Herbert Müller a fourth. As late as 1974 Müller was able to pick up a couple more top-six placings with his 512M.

But there had been one last factory Can-Am effort. At Watkins Glen in 1971 Andretti had appeared with what was basically a 512 spaceframe modified to take a 7-litre engine with revised dimensions and new heads which gave a claimed 680bhp. The new car qualified fifth, which was promising enough. In the race Andretti got to third behind the works McLarens, dropping back to finish fourth. The 712M then went back to Italy and did not race again in 1971, and nor did a

**The Ferrari factory sat 1970 out but returned in 1971 with a new challenger, this one with a 7-litre engine, for Mario Andretti. High wings had been banned by now**

Left **Andretti showed the Ferrari 712M to be a threat to the competition but contested only one 1971 race. A bit of development would surely have turned it into a winner**

Below **Can-Am 712 engine was the biggest ever made by Ferrari. The car itself was not an all-new design, but an adaptation of the 512 run in World Endurance Championship events**

smaller and lighter version which was supposed to appear later in season. Jean-Pierre Jarier did however drive the Andretti car at Elkhart Lake in 1972, finishing fourth, and as late as 1974 Brian Redman was able to hold down third place at Watkins Glen until the ageing car failed.

But the Ferrari Can-Am project, like so many other attempts, was one which might have succeeded rather than one which did. Apart from Porsche, few other Continental manufacturers worried too much about the Can-Am. Matra and Alfa Romeo took part in the Watkins Glen round once or twice when they

**With almost 700 horsepower on tap, Ferrari 712 engine was in the same league as the best American V8s it was racing against, and better than all but the very fastest contenders**

were at the circuit for the world championship race which formed part of the same meeting, and lent some variety to the sights – and especially the sounds – of the race. Milt Minter, better known as a Porsche man, campaigned an Alfa Romeo 33 in several of the 1973 Can-Am races, and picked up a couple of fifth places among the faster machinery.

## A QUIET START FOR PORSCHE

Porsche cars had been taking part in the Can-Am since the first race of 1966, when a privately-entered 906 Carrera had tagged on at the back of the St Jovite field (and finished 14th in the race). In 1969 a works 908 driven by Josef Siffert was among the endurance-racing cars which took part in the Watkins Glen Can-Am race and, in spite of qualifying midfield, finished sixth after a steady drive. He was followed in the next places by four more refugees from the endurance series.

Among them was Englishman Tony Dean, who drove a Porsche 908 of his own, and liked the Can-Am so much he decided to complete the remaining six races in the year. He finished them all, taking a fifth place, a sixth, three sevenths and an eighth.

This was hardly earth-shattering stuff, but Dean returned in 1970 to take in as many of that year's races as he could. His new campaign started with fourth and fifth but there was better to come. At Road Atlanta his underpowered 908 outlasted all the more fancied runners and gave Porsche its first Can-Am win. Dean also took top-ten placings in two of the later rounds and wound up in sixth place on the points table.

In Europe meanwhile the 908 had been replaced as the principal Porsche endurance-racing weapon by the dramatic flat-twelve 917 coupé which, after a bit of sorting, would prove invincible. As soon as it had appeared in Europe, in 1969, its potential as a Can-Am contender was obvious to Californian importers Porsche-Audi, lured by the fact that although the new car's 4.5-litre engine was smaller than the American V8s, its power output exceeded 100bhp per litre, and no stock-blocker could match that.

Porsche-Audi prevailed on their German masters to build a special spyder version, replacing the coupé body with a smooth flowing open shell reminiscent of the 908, though by the end of the season its surfaces had been made flatter (and uglier).

The 917PA first appeared at the 1969 Mid-Ohio round, with Jo Siffert at the wheel. He qualified it with a promising seventh-fastest time, and brought it to the line seventh. He started the next two races from sixth on the grid, blew the engine in one, but took a strong third place in the other. The improvement continued and in practice for the Michigan round Siffert was beaten only by the works McLarens. His season

**Endurance racing star and the first Porsche driver to make a mark in the Can-Am series, Joe Siffert campaigned the 917PA in 1971, paving the way for future German successes**

Above **The Watkins Glen Can-Am race used to be held on the same weekend as the circuit's contribution to the World Endurance Championship. Servoz-Gavin in a WEC Matra 650 in 1969**

Left **Alfa Romeo T33s were another WEC contender seen in occasional Can-Am races. This is Scooter Patrick at Road Atlanta in 1972, but Milt Minter's 1973 campaign was more successful**

continued with two fourth-place finishes and two fifths, and a solid fourth place in the championship.

The 580bhp which the 917PA produced was quite enough to match the opposition at that time, but the car was far too heavy, and the project was abandoned at the end of the year. It was revived in uprated form two years later when Milt Minter drove it for Vasek Polak and earned top-ten finishes in several rounds and sixth place in the championship. Sam Posey was even able to achieve a fifth place with it in one 1972 round.

Dean's efforts with his 908 aside, the only Porsche presence in 1970 had been at Watkins Glen, where several 917s entered in the world title event also took part in the Can-Am round. Siffert in the fastest of them battled for second place with Gurney's McLaren and Revson's Lola, and crossed the finish-line with only Hulme ahead.

What's more, Richard Attwood (winner at Le Mans that year) and Vic Elford brought their similar cars home third and fourth, and Gijs van Lennep and Brian Redman in yet more 917s were sixth and seventh.

Thus encouraged, the factory launched a renewed attack on the Can-Am series in 1971. A new tubular aluminium space-frame was designed to take a more powerful 5-litre engine now capable of around 630bhp. The 917/10, as the new car was called, was equipped with a stubby body which used tail-fins but no aerofoil. The liberal use of magnesium and titanium in the car's construction made it lighter than the PA, but it was still heavier than the latest McLarens.

It made its debut in practice for the Watkins Glen round, but Siffert couldn't get it around the track as fast as van Lennep's normal 917 coupé. But as the race progressed the Swiss ace went faster and faster, and at the end was third behind the works McLarens. Van Lennep was ninth.

Left **First serious Porsche Can-Am car was the 917PA, an open version of the new 917 WEC car. Jo Siffert, here at Riverside, took it to fourth place in the 1969 series**

Below **Porsche, like Matra and Alfa Romeo, had first contested Can-Am races with WEC cars. This is Siffert at the wheel of a 908 in the 1969 Watkins Glen qualifier**

Siffert followed this up with second places in the next two rounds, and completed the campaign with a fourth and two fifths, to repeat his 1969 fourth place in the standings.

During the season a more powerful (650bhp plus) engine of 5.4 litres was used, and there were rumours of even more for 1972. The secret weapon being whispered about was turbocharging. This was firmly established on the ovals, and the winning car at Indianapolis had been so powered virtually every year since 1968. Now Porsche were believed to be investigating the technology with future Can-Am racing in mind.

## TRUCK TURBOS

Back in Germany Porsche built a special flat-16 engine of 7.2 litres which reportedly produced 880bhp, but opted instead for the familiar 5-litre flat-12 coupled up to a pair of exhaust-driven turbochargers designed for use on commercial vehicles. Running at 1.3 bar, these gave some 900bhp. But more significantly, instead of Porsche torque being about two-thirds of that available to the best Chevrolet V8s, it was now about 20% better.

To take the extra power a new car was built with a strengthened aluminium-tube frame and stronger and heavier gearbox and brakes. The body was distinguished by a 'snow-shovel' nose and a massive rear wing and although the car was still by

Right **The 917PA was powerful, but too heavy to be truly competitive, and the factory abandoned the project after the 1969 season. Siffert at Riverside again**

Above **Siffert (right) and works mechanics with the prototype 917/10 before the 1971 series. The car was much lighter than 917PA, with new chassis and more powerful engine**

no means a lightweight, its improved power and torque gave it an advantage over the opposition. The model was known in America as the 910/10K (for Kompressor, or supercharger) but European observers, more used to Porsche applying that suffix to distinguish a short (kurz) car from a long one, preferred to describe the new Can-Am contender as the 917/10T.

It had already been decided that Roger Penske would run the factory's Can-Am programme, and his regular driver Mark Donohue had in fact spent many weeks on the project in Germany. His input as an engineer and test driver was a major component in the success the 917/10T would enjoy. In the midst of this Donohue found time to nip back to the States and win the Indianapolis 500 in a Penske-entered McLaren: to add to the three Trans-Am championships he had now won, as well as his two earlier USRRC titles.

The Can-Am series suffered a major blow before the season started when Johnson Wax, main sponsors from the start, announced their withdrawal. Nevertheless, there were nine races on the schedule.

First time out, in the Mosport series-opener, Donohue qualified the 917/10T on pole, relegating the works McLarens to support roles. In the race a stop to fix a broken throttle-linkage cost him three laps, but he came back to finish second.

Above **Siffert debuts the 917/10 at Watkins Glen 1971. In spite of its improved power-to-weight ratio, the spyder was still no match for the Anglo-American opposition**

Right **Pedro Rodriguez in a WEC Porsche 917 coupé in the 1970 Watkins Glen Can-Am race. A similar car finished second, prompting revival of plans for a dedicated Can-Am car**

But he then crashed heavily when the car flipped in practice for Round 2 at Road Atlanta. Donohue damaged a knee, and was told he would not race again that year.

Penske signed George Follmer, who had just won the Trans-Am championship, to stand in, and the new combination was unbeatable. Follmer in the Penske 917/10T won at Road Atlanta, Mid-Ohio (where he started from pole) and Elkhart Lake, though at Watkins Glen the car did not perform to its usual level and he had to make do with fifth.

Donohue returned to the fray for the Mid-Ohio round, so Penske fielded two cars from then on, as Follmer had a good mathematical chance of taking the title. Donohue duly qualified on pole at Mid-Ohio with Follmer second and they ran 1-2 until a tyre blew on Donohue's car, and he left the road, and Follmer ran out of fuel, rejoining to place fourth. Donohue then won at Edmonton while Follmer took the Laguna Seca

**Body of 917PA was similar to 908 spyder, and one of the smoothest of all 1969 Can-Am cars. As the season progressed however lines were spoiled by addition of aerodynamic aids**

**Turbocharged versions of the 917/10 finally brought an end to McLaren domination of the Can-Am in 1972. This is Follmer's championship-winning car in its 1973 colours**

and Riverside rounds; between them they also took a second place and two thirds in these races. This gave Follmer the championship while Donohue, in spite of missing so many races, occupied the fourth rung of the ladder.

Minter in a normally-aspirated 917/10 enjoyed a remarkably consistent season, finishing seven of the nine rounds, six of them in the first six places. His best finishes were a second and two thirds, and he was third on the points table. In a similar car long-time US Porsche exponent Peter Gregg logged up a tally of four top-six finishes, with his best result a third.

## DRIVING A PORSCHE, OR NOWHERE

After the 1972 season the Can-Am series went into decline. McLaren and Lola were no longer building cars for it, and Chaparral were long gone. And the reduced sponsorship now available meant that only eight circuits agreed to stage qualifying rounds.

There were still one or two serious American challengers, but nothing seemed to stand in the way of another Porsche walkover in 1973. The 917/10Ts appeared unbeatable, but the

**In spite of running a non-turbocharged engine, Milt Minter in his 917/10, here at Watkins Glen, picked up a string of places to finish third in the 1972 series**

Germans had something even better on the drawing-board, a revised version called the 917/30. The new car was seven inches longer with streamlining improved in consultation with a specialist company in France. It was heavier than its predecessor, but larger turbochargers meant an increase in

Above **George Follmer won the 1972 series in a Penske team turbo Porsche 917/10, and was second in the same car, now run by the Bobby Rinzler team, the following year**

Left **Star of the 1973 series, Mark Donohue in the Penske team Sunoco Porsche 917/30. Turbocharged engine produced 1100bhp, the most powerful seen on road circuits anywhere**

power output to a fantastic 1100bhp at 1.3 bar. This made it by far the most powerful car ever seen in road racing, with twice the output of a contemporary Formula One engine.

Only one 917/30 was run, entered by the Penske team. Donohue was again the driver, and qualified fastest for every single round. He retired at Mosport after a collision with another car, then finished second at Road America, where he was slowed by an oil-leak. But from then on he was unbeaten, winning at Watkins Glen, Mid-Ohio, Elkhart Lake, Edmonton, Laguna Seca and Riverside. His points total was more than double that of his closest challenger.

The back-up was no longer provided now by McLaren and Lola, but by other Porsches: if you weren't driving one of the German cars in 1973 you weren't in the picture. Fastest of the several 917/10T cars taking part this year was usually the Vasek Polak entry driven by Jody Scheckter, but it was also one of the least reliable. The young South African did however take a second place and two thirds. The ex-Penske 917/10Ts were run by the Bobby Rinzler team, with Follmer and Charlie Kemp as drivers. They each won one of the opening rounds while the 917/30 was being sorted out, and Follmer also took two seconds and a third in the series while Kemp, hitherto a midfield runner in Lolas and McLarens, managed another third and a fourth. Hurley Haywood's tally in a privately-entered 917/10T was one second, two thirds and a fifth, while Hans Weidmer took a second in another. Steve Durst raced a non-turbocharged 917/10 and picked up fifth place in the first round while Brian Redman contested a couple of other rounds in Polak's further-developed 917PA, without finishing. Follmer, Haywood and Kemp filled the places behind Donohue in the championship.

## TURBOS EFFECTIVELY BANNED

The SCCA had for the past few years been running a very successful series for Formula 5000 single-seaters, which were rather like Formula 1 Grand Prix cars except that they used American production-based V8 engines of five litres. The club was keen to standardise its main professional categories, and proposed that Can-Am cars be limited to the same engines. The individual race-organisers were unanimously against the proposal, and it seemed the status quo would remain. But then America, along with the rest of the world, was hit by the

Above **Future world champion Jody Scheckter was among numerous European-based drivers who used Can-Am series as a stepping-stone, here in Porsche 910/10 turbo in 1973**

Right **Scheckter follows the pace-car at the start of a 1973 race with Donohue's 917/30 alongside. Haywood in another 917/10 and Hobbs in his ex-works McLaren lead the chase**

international oil crisis. The turbocharged cars were prodigious consumers of fuel, and although the SCCA did not ban them from the 1974 Can-Am, they did build a fuel-consumption element into the rules, and this had almost the same effect.

This result was the least successful Can-Am series ever. Penske brought the 917/30 out for only one round, at Mid-Ohio, where it had been most economical in 1973. By running on reduced boost it was hoped it might last the distance. Donohue had now retired from race-driving, though he would be back behind the wheel before the year was out – only to die after a freak accident in the following year's Austrian Grand Prix. Redman was the 917/30 driver at Mid-Ohio, and duly qualified on pole, but tyre trouble left him second at the end. Haywood in his own 917/10, running unturbocharged, was third, and also ran at Elkhart Lake, but did not finish on that occasion. And that was the end of the Can-Am Porsche story.

CHAPTER EIGHT **CAN-AM**

# America after CHAPARRAL

The make that won the low-key 1974 Can-Am Challenge was an American Shadow, for participation by 100% local efforts had not stopped when Chaparral withdrew from the fray after the 1970 season.

Indeed, there had always been smaller teams, most making little impact on the series as a whole. Many names appeared on the grid but fleetingly, then disappeared. Most of them were one-off specials, though McKee was one make which was produced in quantity, and Genie was another. Both these names graced the grids for several seasons. Less permanent US and Canadian features of the scene over the years included Webster, Mirage, Burnett, Chinook, Wolverine, Hamill, Stanton, Platypus, BVC, Moore, Rattenbury and Sting.

There had been some major American players over the years, too, notably the Ford Motor Company. The Detroit giant had become heavily involved in other forms of motor racing by the mid 1960s, at Indianapolis with their DOHC engine, at Le Mans with the GT40 project and subsequently in Formula One with the Cosworth DFV engine, and were multiple winners in every category. It must therefore have been galling for them to see the success that the engines of their arch-rivals Chevrolet were enjoying in the Can-Am series each year.

Ford never made a high-profile foray into the series, but encouraged private teams to run Ford engines in proprietary chassis. We have already seen that a variety of Ford-powered McLarens and Lolas had appeared in the early years of the series, but similar engines were seen in a variety of US chassis as well.

Early in the life of the Can-Am the Ford Motor Company tested a couple of prototypes of their own, based on the MkIV development of the GT40. But the Calliope project, as it was called, was abandoned without the car being raced.

Instead they channelled funding to a variety of independent projects. Carroll Shelby had long been a good Ford customer, and for 1967 had Len Terry in England design a new King Cobra to make use of Ford power. The car's aluminium monocoque originally featured transverse coil springs, though these were replaced after testing. It appeared only once, at the end of the season, but did not handle and Shelby did not persevere with the project. The car turned up again in private ownership three years later, but changing its name to the Terry T10 had not made it any more competitive.

A second Ford-backed car had also appeared in 1967. This one was also designed and built in England, by Len Bailey and Alan Mann Racing, with a conventional monocoque. It was powered by the then-new 5.8-litre Ford engine which, in fuel-injected form, was probably good for 550bhp. Called the Honker II, this car was entered by the famed stock-car team of Holman and Moody in a number of 1967 races, with Mario Andretti as driver. By the Riverside round he was able to qualify it in a respectable fifth position, and held a strong fourth place in the early stages of the race. The Honker retired after ten laps however, and had further problems in practice at Las Vegas. And that was the last that was seen of the car.

Two years later Holman and Moody were responsible for another Ford-powered project built by Alan Mann. Called the

**Original Shadow was little more than a tiny-wheeled skateboard with huge engine. The theory was right, but in practice the car worked better the bigger it became**

AMR CA/Ford, or the "Open Sports Ford", it contested only the last couple of rounds of the 1969 series. British-based Australian Frank Gardner, then a big name in touring-car, F5000 and sportscar racing in Europe, showed reasonably well with it in the Riverside race that year. Brabham took the car over for the Texas round, and came home in a solid third place.

At the same time Ford gave their backing to the Agapiou brothers, who had acquired the two Calliope cars. Entered now as a Ford G7A, and using a conventional big-block Ford engine, one car appeared in several 1969 and 1970 rounds, but in spite of having drivers of the calibre of Brabham, Revson, Elford, Hobbs and Follmer, it didn't make much impression.

After all that, Ford washed their hands of the Can-Am series, and basked instead in the success of the their engines in Grand Prix and Indycar racing.

## ONE-OFFS

Among the smaller American outfits which had a go in the Can-Am series, several tried to out-Chaparral Chaparral in complexity. One of the most bizarre attempts to get to grips with the front-runners was a four-wheel-drive car called the Hoare Mac's It Special, which had one 800cc twin-cylinder Rotax snowmobile engine to drive each wheel. It turned out in practice for one 1970 race but never actually raced. The minuscule AVS Shadow of the same year ran on tiny wheels but worked better on paper than in practice, and the same is true of the Costello SP8, which came out of Canada three years later.

**By 1973 Shadow layout had become conventional, though this DN2 model did run on occasion with turbocharged engines. Either way, it was no match for the turbo Porsches**

Above **The Shadow Mk2 remained smaller than the opposition, and showed great potential. Jackie Oliver took one to an excellent third place at Edmonton in 1971**

Right **By 1973 the Shadow operation had moved to England, where the DN2 was designed and built. Long, wide monocoque carried bodywork with distinctive shovel nose**

Its guiding principles were minimum size and weight, to which end it was powered by a small-block Oldsmobile engine.

An earlier Can-Am oddball had been the Caldwell D7 of 1967/68. Designed to take advantage of the latest fifteen-inch tyres, the car was equipped with De Dion-type suspension front and rear, and with one of the first chassis-mounted adjustable wings to see the light of day. This made the car unstable however, and was soon taken off, but even the replacement of the monocoque with one made of magnesium, and the enthusiasm of its backer/driver Sam Posey, did not turn the car into a front-runner. And it was an American car which first ran a turbocharged engine in the Can-Am, when a McKee powered by a 6.4-litre iron-block Oldsmobile managed an eighth place in one 1969 race. The car was then dramatically updated with a 7.5-litre engine, automatic transmission, four-wheel drive and an air-brake, but it was too complicated for the small operation, and after appearing in practice for one race, was put away for lack of funds.

Any of these cars might have made a more permanent mark had development been possible, but as the whole idea of their novel approaches was, in most cases, to find a cheap way of matching the winners, funds were simply not available to take the projects further. We can laugh at some of their ideas now, but that's the way people greeted some of Chaparral's ideas when they first appeared. Yet within a few short years top-level racing cars without wings and ground-effect were unthinkable.

Hertz
L&M
LOLA
1
GOODYEAR
Castrol

PORSCHE + AUDI
SUNOCO
DX
PORSCHE
+AUDI
SUNOCO
PORSCHE
+AUDI

DX
DieHard
6
PORSCHE
+AUDI
SUNOCO

Page 81 **Frantic pre-race activity as the cars are readied for the 1971 Watkins Glen race. No. 1 is the Lola T260, partly obscured by Tony Adamowicz's ex-works McLaren M8B**

Opposite, top **The ultimate Can-Am car. Porsche's 1000-horsepower turbocharged 917/30 pushed performance barriers to their limit. This example is preserved in Porsche's own museum**

Opposite, below **After hiccups in the first two rounds, Mark Donohue in the Penske Racing Porsche 917/30 won all six of the remaining rounds in the 1973 championship**

Above **McLaren and Lola built 'production' cars for private teams. McLaren M6B of 1968 was one of most popular. Later owners of this example have incorporated newer features**

Right **Regardless of the factory's choice, private owners were able to fit their own engines. This McLaren M6B has a retrofitted larger engine with non-period fuel-injection trumpets**

Overleaf **The Bruce and Denny Show on canvas; artist Michael Turner's evocative painting of McLaren leading Hulme in M8Bs at Michigan in 1969. Michael Turner was himself in the thick of the Can-Am action, and produced several unforgettable images from those few unforgettable seasons**

GOOD YEAR
4

REYNOLDS
GOODYEAR
5
1
4
Gulf
MICHAEL TURNER '70

REYNOLDS
ALUMINUM
6
McLAREN CARS
Gulf
BOSCH
GOOD YEAR
REYNOLDS
ALUMINUM
Gulf
EAGLE
GOOD YEAR

EAGLE

1966 Chaparral 2E
(See commentaries, p112)

1968 McLaren M8A

1969 McLaren M8B

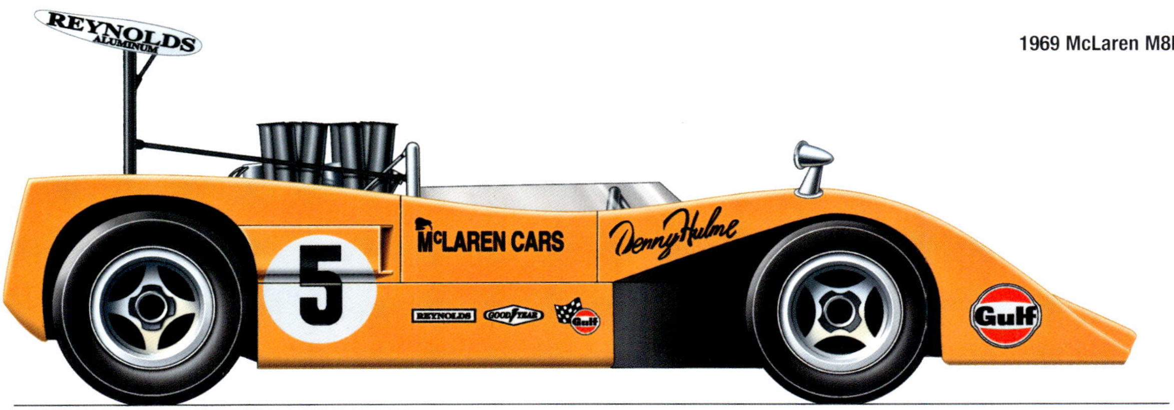


Opposite **McLaren M8D was the last Can-Am car produced by Bruce McLaren before his tragic accident in 1970. Engine size had by now grown to 8 litres, giving around 700bhp**

1971 Lola T260

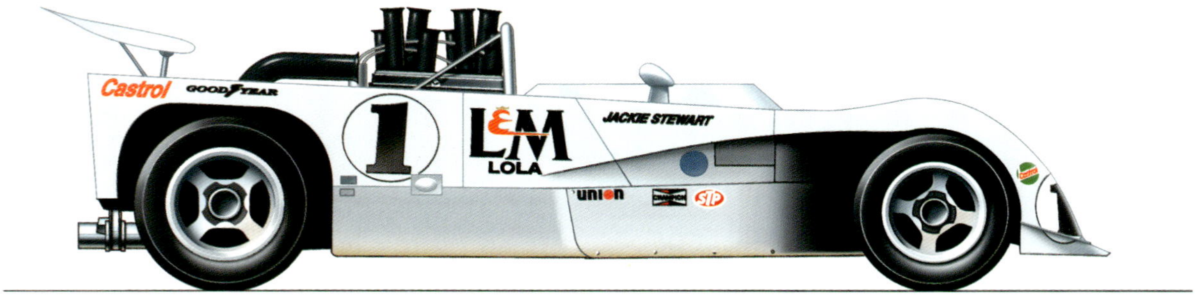


1972 Porsche 917/10 Turbo

1973 Porsche 917/30

1974 Shadow DN4

M8F was the last McLaren to dominate the Can-Am series, but was seriously threatened by Lola and Porsche in 1971. The following year turbocharging tipped the balance in Porsche's favour

PORSCHE WINS RIVERSIDE CAN-AM
1. DONOHUE PORSCHE 917-30 TURBO
2. HAYWOOD PORSCHE 917-10 TURBO
3. KEMP PORSCHE 917-10 TURBO
4. NAGEL LOLA T260
5. MINTER ALFA ROMEO T33/3
6. PECKHAM McLAREN M8C
SUNOCO
PORSCHE+AUDI
GOODYEAR
DIE HARD
PORSCHE

Opposite **US Porsche distributors were quick to cash in on publicity created by their 1973 Can-Am successes**

Right **Ferrari never seemed to take the Can-Am seriously enough. Cars made odd appearances as late as 1974, when Redman drove this 712M without success in the Watkins Glen round**

Below **Lola T260 driven by Jackie Stewart in 1971 took over Chaparral's role of threatening the McLarens. The following year's contender was not so successful, however**

REYNOLDS
5
McLAREN CARS
Gulf

Valvoline
RC
CHAMPION
Charlie Kemp
23
PORSCHE
+AUDI
union

Opposite, top **McLaren driver Denny Hulme, the 1967 Formula One World Champion, also took two Can-Am titles and was runner-up on three other occasions, including 1969 (pictured)**

Opposite, bottom **While Donohue steamrollered to 1973 championship success in the Porsche 917/30, the competition had to make do with the older 917/10 Turbo. Charlie Kemp was fourth in the series, behind two other 917/10s**

Above **An earlier example of those sporadic Ferrari efforts which brought colour and noise to Can-Am: in this 1969 race, at Watkins Glen, Amon beat everyone except the uncatchable McLarens**

Right **Denny Hulme accepts the plaudits of onlookers after winning at Watkins Glen. He was successful here in both 1970 and 1972**

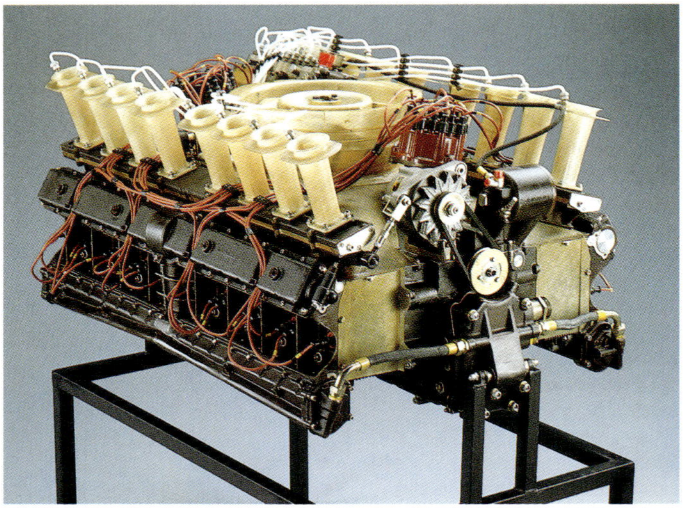

Top **Jacques Villeneuve, uncle of the later Grand Prix driver of the same name, won the 1983 'postscript' championship at the wheel of a Frissbee**

Left **Without the converted F5000 Lolas there would have been no Can-Am series revival in 1977. This is Don Breidenbach's example, in action at Laguna Seca**

Above **Porsche built a 16-cylinder engine for the 1972 Can-Am series but didn't race it, preferring to turbocharge the tried and proven flat-12 instead**

Above **The M8D model brought mixed fortunes to the McLaren team in 1970. McLaren was killed testing the prototype and Hulme crashed another, but won the championship**

Right **The office. M8D cockpit has provision for a passenger, not for practical purposes, but to meet the rules**

Overleaf **The Porsche successes continued, and with them came more publicity opportunities for Porsche-Audi, the US distributors**

PORSCHE + AUDI
GOODYEAR
SUNOCO
DIEHARD
PORSCHE WINS
LAGUNA
SECA
CAN-AM
1. DONOHUE PORSCHE 917-30 TURBO
2. OLIVER SHADOW MK2
3. HAYWOOD PORSCHE 917-10 TURBO
4. BROWN McLAREN M8F
PORSCHE

Without the Can-Am's most important rule – There Will Be No Rules – the opportunity to try out such innovations in international competition would not have been possible.

Less dramatic than some of these ideas, though still departing from basic principles, was the Autocoast Ti22. Designed by English-born Peter Bryant, its full-length monocoque was made from titanium, whose chemical symbol (Ti) and atomic number (22) gave the car its name. The body was a startling wedge shape trimmed with full-length fences.

Fitted with an alloy big-block Chevy engine and Hewland LG600 gearbox, and driven by British F1 driver (and 1969 Le Mans winner) Jackie Oliver, the Ti22 appeared in the last three races of the 1969 series. It arrived at the first of these, Laguna Seca, untested and had to start from the back of the grid. It finished 13th.

The lessons learned here became obvious when Oliver qualified fourth at Riverside, and was battling for third place in the race when the car retired. He was in the same position at Las Vegas when the car once again went out.

Such promise meant the Autocoast Ti22 was taken very seriously when it returned in 1970. Oliver underlined its form by qualifying third for both opening races of the series, with only the works McLarens faster. In the Mosport race he ran ahead of both McLarens at one time or another, finally finishing second. But on the first lap of the second round the car was caught on St Jovite's notorious hump, flipped onto its back, and was destroyed.

A revised version of the car was quickly built as a replacement and, backed now by TRC rather than Autocoast, appeared in the last two rounds of the year. Oliver hounded Hulme's McLaren for the last 40 laps of the Laguna Seca race, finished just over a second behind at the end, and went on to take another second at Riverside. The points he gained during the season were enough to leave him fifth in the championship.

But the project was then sold, and the new owners did nothing about racing it until, after another change of hands, the car reappeared at the end of the 1971 season. David Hobbs was now the driver and showed the Ti22 was as competitive as ever by qualifying third for the Laguna Seca race. He retired there after a collision with a back-marker, and then went out at Riverside as well. Various problems in practice there had meant the car started from the back of the grid, but Hobbs fought his way up through the field in great style and got up to fifth place before damaging the bodywork on a marker-tyre.

It appeared one final time, in amateur hands, at the same track a year later, but made no impression.

### SHADOW

The list of new American cars which appeared in 1969 and 1970 contained only one name which would last, and then only after abandoning its individuality in favour of a more conventional approach. That make was Shadow. The 'tiny tyre' prototype was built by Advanced Vehicle Systems to a design by Trevor Harris on attaining the principle of the smallest possible frontal area. To achieve this, 10in wheels were required, and the project went ahead once Firestone agreed to develop special-sized tyres for it. The resultant car was a mere 17 inches from the ground at the front and 19 at the rear. This was seven whole inches lower than the contemporary McLaren, and gave a frontal area around two-thirds the size. In fact it looked nothing more than a slightly overgrown go-kart with bodywork and a huge engine in the back.

The original AVS Shadow Mk1 was built around a very stiff anodised aluminium monocoque with suspension by parallel links containing four small springs on each side at the front. The rear springs were mounted horizontally, though this layout was later changed to a similar system to the front's. Brakes were mounted inboard at the rear, and there was also provision for an air-brake, though this was outlawed by the ban on 'movable aerodynamic devices' which came in before the car raced.

The car was scheduled to race in 1969, but problems showed up with the aerodynamics and, although the design had paid a lot of attention to achieving adequate engine-cooling, this needed further work too. As a result, the revolutionary car did not race until 1970.

During that season it proved itself capable of qualifying in the top half-dozen, but neither of its drivers, Follmer and Elford, could get it to the finish-line. The races it ran showed that the aerodynamic and cooling problems were still there, and the tiny tyres made it very hard work to drive.

And so after just three race AVS boss Don Nichols withdrew from the series to devise an improved version for 1971. Design of the new car was entrusted to Peter Bryant, who retained the principle of small wheels, though they were bigger than the Mk1's. Also retained was a low-profile body similar to the old car's, but with aerodynamic fences on the top of the body-sides and a low-mounted rear wing. The car's basis was a compact aluminium semi-monocoque with the brakes mounted inboard at the front but outboard at the rear. A Chaparral-built Chevrolet engine drove through a four-speed Weismann gearbox.

Now painted a sinister black in honour of the team sponsor, the UOP oil company, and with Oliver as driver, the Shadow Mk2 proved fast but unreliable. It was usually one of the first four or five in qualifying, and once or twice got higher than that in a race, but its only points finish from eight starts was third place at Edmonton.

The Mk3 at last gave up on smaller than normal tyres, but was otherwise more or less the same as the Mk2. One departure was the side-mounting of the radiators, and the use of a wing across the nose, and the whole package was lighter than the older model. During the season engine man Lee Muir worked hard on a turbocharged version of the Chevy V8, from which

**Rules for 1977 restricted the amount of fuel that could be carried, so the new Shadow DN4 was smaller. It was also ready for a pre-season testing programme, here at Silverstone**

1200bhp was expected, but it did not appear until the last round, and then blew up in practice. The team also experienced transmission problems during the year, and swapped the Weismann gearbox for a Hewland.

Competition was tougher now, but throughout 1972 Oliver usually qualified the car in the top six places, and took second place at Mid-Ohio, backed by a third and a couple of fourths.

The following year Nichols moved the Shadow operation to England, as he was embarking on a Formula One programme and a European base made sense for that. His F1 designer, Tony Southgate, was given the task of devising a new Can-Am car to take the turbo Chevy engine, and came up with a full-length aluminium monocoque which was longer (by five inches) and wider than the Mk3, its body distinguished by a shovel nose. The Shadow DN2 was, apart from the Porsche 910/30, the only new car for the 1973 series.

The turbo engine was not ready when the season began, and a normally-aspirated 8.1-litre Reynolds-built Chevy unit was used instead. It produced 735bhp but this was no match for the turbo Porsches, now in their second year. Shadow did run their turbo Chevy three times during 1973, but it never went well, In both forms the DN2 handled poorly, and with the team concentrating on F1, there was little time to develop the Can-Am car.

In view of all this, Oliver did well to qualify in the top five for three races, and to bring home a second place and a third.

### DN4 DOMINANCE IN FINAL YEAR

The Shadow DN4 for the 1974 series was built to the new maximum fuel allowance (37 gallons) so was more compact than its predecessor; otherwise its layout was basically the same as the DN2's. It retained the 8.1-litre Chevy engine, running without turbo, of course, but probably producing 800bhp.

Nichols took out insurance against possible failure by running two cars, Follmer rejoining the team to run alongside Oliver, his team-mate in the Shadow F1 team. Significantly, in the tradition of McLaren and Porsche, Shadow's cars were ready in time for a comprehensive pre-season test programme.

The DN4 was the car which swept the 1974 series, its domination being as complete as those of McLaren and Porsche in earlier years. On the other hand the lack of public interest, not to mention so little in the way of competition, made the Shadow success rather hollow. The black cars qualified fastest and second fastest for four of the first five rounds, being beaten by the Porsche 910/30 on the one occasion it appeared. Oliver was faster in the first round but Follmer outqualified him in all the others. At Mosport, Road America and Watkins Glen they finished 1-2, with the Englishman ahead each time, and Oliver also won at Mid-Ohio. Follmer was a retirement there however, and neither car finished the fifth race, at Elkhart Lake.

Eight races had been scheduled, though one organiser pulled out after he'd seen the first race. Now, immediately after Elkhart Lake, the last two rounds were also cancelled, leaving just the five rounds already held to decide the series. No Shadow driver had yet finished higher than eighth on the Can-Am points scoreboard, but at the end of the abbreviated 1974 series Oliver had a clear lead on points, and Follmer was next.

The 1974 Can-Am formula was supposed to have been an interim step before the SCCA's 5-litre limit came in for 1975. It was fairly obvious however that circuit support for the class was almost non-existent, and when no likely constructors indicated any interest, the club had no option but to call it off.

The SCCA announcement suggested that rising costs had killed the series, and expressed surprise that entrants were not attracted by the apparent saving to be made by restricting engine sizes. After it was all over, some blamed the years of domination by one make – first McLaren and then Porsche – for killing off the Can-Am. Others said the cause was the sudden change of rules which led to the departure of Porsche and therefore of a glamorous major player. Others still, probably more accurately, believed the series to have been doomed from the moment the rule-makers started intervening.

The truth of the matter however is that the series had simply run its course. When the Can-Am series started, Group 7 racing was equally strong in Britain and North America. But after the 1966 season the trade sponsors in Britain withdrew their support, and the class survived on that side of the Atlantic only at low-key club meetings. Interest in the formula never died however, and as the Can-Am series went from strength to strength in North America, there was talk of bringing big-banger sportscar racing back to Europe.

In 1970 a series of pilot races was run on the Continent, as a prelude to the launch of the so-called 'Interserie' the following year. Privately-entered Porsche 917 spyders and McLarens were seen in regular action, with the occasional works sortie by BRM or March. The increasingly German-centred series was was never well supported however, and the Interserie lurched to its end about the same time as the original Can-Am.

**With no turbo Porsches to worry about, Shadow DN4s dominated the 1974 Can-Am, Oliver winning the title with Follmer runner-up. It was the last year of the original Can-Am series**

# EPILOGUE

Although the 'Golden Age' of the Can-Am was over. the name was revived after a couple of years' break, and in fact continued for another ten years after that. The new series was however full of the sort of rules and regulations whose very absence had made the original series such a success, and was contested by few teams other than US-based operations. Lola and March both built cars for it, but there was little direct factory involvement.

In another link with the past, Formula One drivers took part in most years, but in this series only Jacky Ickx was a real star of the calibre of the world champions who had regularly graced the grids of 1960s Can-Am races. On the other hand the series did, over the years, attract a number of rising stars, including future world champions Alan Jones and Keke Rosberg and Canadian superstar Gilles Villeneuve. But the names that came back year after year were, by and large, the sort of North American amateurs who had filled the rear two-thirds of the grid in the 1966/74 series.

Within a year or two of the death of the old Can-Am series in 1974, the SCCA's Formula 5000 championship began showing signs of going the same way, so for 1977 the club announced a replacement. The new series would continue to cater for F5000 cars, but they must now run with full-width bodywork, and under the revived Can-Am name. The SCCA thereby finally achieved its ambition, ill-considered though it may have been, of running a 5-litre 'stock block' Can-Am series, though few could see quite what the point was of a central-seat sportscar series.

Engines for the new series would be limited to five litres, though the F5000 provision for dedicated racing units of three litres remained. In practice virtually everyone used Chevrolet

**Cars in the revived 1977 series were basically rebodied F5000 single-seaters, hence central-seat layout. This is Patrick Tambay's Lola 333CS at the Laguna Seca round**

The team run by Carl Haas and Jim Hall was the one to beat in the early years of the revived series. Patrick Tambay, here at Sears Point, won 1980 series for the team

V8 engines, in this case developments of the 5-litre Camaro unit.

In spite of many detractors, the revived series had one significant supporter, Carl Haas. Still the US distributor for Lola cars, Haas arranged with the factory to convert a number of F5000 cars for the new formula - which in practice meant little more than the fitting of wider bodywork - and offered them for sale to teams wishing to enter the 1977 Can-Am series. Without Carl Haas, there would not have been a series in 1977, for few cars other than Lolas contested the nine races which made up the championship that year.

### TOP-CLASS DRIVERS

Haas had been entering Lolas in US F5000 races for several years in partnership with Chaparral boss Jim Hall, with considerable success. Indeed, their driver Brian Redman, having now turned his back on European racing, had won the US single-seater title for Hall/Haas in each of the last three years.

The Hall-Haas team entered a Chevrolet-powered Lola T333CS for Redman in the 1977 Can-Am, only to have their plans disrupted by a bad crash in practice for the opening round. The injuries Redman suffered put him out of racing for the rest of the year but Haas found a suitable replacement in Formula Two star Patrick Tambay. The young Frenchman took part in seven races and won six of them, finishing fourth in his other start. He was a clear winner of the championship.

For 1978 the team hired Australian Alan Jones, then in his fourth season of Formula One and his first with the Williams team. His Grand Prix commitments meant he missed one of the ten Can-Am rounds, but he qualified on pole for all the others, winning five of them and taking a second place and a third as well to give the team another championship.

Belgian Jacky Ickx, in his last year of Formula One and already a three-time winner of the Le Mans 24hr race, was the Carl Haas Racing driver in 1979, winning five races and the title. Jones was brought back for one race, when Ickx was committed to Le Mans, and added that event to the team tally for the year.

Throughout the first three years of the revived series Haas and other main Lola entrants relied on the T333CS model, but for 1980 the factory produced a dedicated central-seat Can-Am model in the form of the T530. Patrick Tambay, now a member of the McLaren F1 team, returned to the series to drive the Haas entry and, with wins in six of the ten rounds, took his second Can-Am title.

But after just four years the series was already showing signs of collapse. A comparative lack of technical innovation, coupled with competition from an increasing number of other North American championships, meant it had never attained the popularity of its predecessor. Spectator attendances were poor, and fields were getting smaller. Jim Hall had now left to pursue a career as an entrant and constructor in Indycar racing, but Haas contested one last Can-Am season in 1981. His driver this year was Jeff Wood, one of the stars of US Formula Atlantic racing, but although he took several placings he did not win any races, and finished up in fifth place on the championship table.

Team VDS driver Michael Roe was the pace-setter in 1983 (above) and 1984. Unreliability cost him the championship the first year, but he made short work of the 1984 title

## TEAM VDS

The second most successful team from the outset of the revived series had been that of Belgian Count Rudi van der Straten, well known as an entrant in European sportscar and F5000 races. Retired F1 driver and one-time works McLaren Can-Am team-member Peter Gethin drove the Team VDS Lola T333CS in 1977 and, although he won only one round, he finished in the top six places in no fewer than eight – including five in the top three – and was runner-up on the final points table in what was to be his last year of race driving.

The following year Australian Warwick Brown, a former Tasman F5000 champ and a US F5000 regular, achieved an even better record – one win, five second places and two thirds – but again left the team second to Haas in the standings. F1 new boy Geoff Lees joined Team VDS in 1979 and Geoff Brabham, son of Sir Jack, the following year, each winding up third in his championship campaign.

Brabham had however managed to win one round, and at the wheel of the VDS Lola T530 in his second year with the team, 1981, he won another. Work was meanwhile under way on the team's own contender, and Brabham gave the VDS 001, as it was called, its maiden victory later in the year. In addition he took three second places and three thirds to give Team VDS its first Can-Am championship title.

Series regular Al Holbert, twice champion in the rival IMSA series and now a Can-Am regular, was the VDS driver in 1982, and finished in the top three placings in seven races. Four of these placings were firsts, but that was good enough only for the runner-up position in a championship which was growing less and less important.

Indeed, in 1983 only six circuits could be persuaded to stage rounds in the championship, and Team VDS was one of the few serious contenders. Rising Irish star Michael Roe was the driver this year, but although fast, the VDS also proved unreliable, and managed to pick up only a few placings.

The dying series was given a shot in the arm in 1984 under the direction of new promoter Don Walker, and the number of qualifying rounds returned to ten. He also entered the VDS 002 for Roe, who took the title with victory in seven of those rounds.

Although Lola was the big name in the early years of the revived series, both in numbers and in terms of success, there were other contenders too. As early as 1977 Canadian Walter Wolf ran a car for a young countryman called Gilles Villeneuve, but after a single third place the team pulled out of the series. Alan Jones ran a Shadow DN6 that year, powered not by the almost universal Chevrolet engine, but by a Dodge. But this

effort was even less successful than the Wolf. Australian manufacturer Elfin entered cars for F1 and USAC driver Vern Schuppan in 1978 and 1979, but without much success. Another antipodean contender in the latter year was the McRae GM3, built in New Zealand. Its constructor/driver Graham McRae was a past champion of both the European and American F5000 series, but his Can-Am car was a disappointment. The rest of the series produced a number of American-built challengers, most of them based on the ubiquitous Lola. Former Can-Am champion George Follmer had been among the second-rank Lola drivers in 1977, but for the following year had his car modified and renamed the Prophet. With this he finished fifth on the points table, in spite of a bad crash late in the year. The Prophet was fifth again over the next two years as well, this time in the hands of Formula Atlantic, F2 and endurance racing driver Bobby Rahal, later of course a star of Indycar racing.

Like Follmer, F5000 regular Elliott Forbes-Robinson had campaigned a private Lola in 1977. He ended the season third, but, again like Follmer, had his own developments carried out for 1978. In this car, called the Spyder NF10, EFR won two races, and fourth place in the championship.

Movie star and amateur race driver Paul Newman took over the Spyder project for 1979, and with the Finn Keke Rosberg, then in his second year of F1, as his driver, produced the surprise of the season. The Newman/Rosberg/Spyder NF11 combination was the fastest, starting eight races from pole, leading Ickx in the Haas Lola. Rosberg won the first round but from then on suffered several mechanical failures and crashes and could add only one further victory to his tally, which left him fourth in the points standings at the end of the year. Newman also entered a second NF11 for Forbes-Robinson who

Below **First VDS Can-Am title had fallen to Geoff Brabham in 1981, driving the team's Lola T530, here at Watkins Glen, and later the VDS-001. Chasing him is Fabi's works March 817**

Overleaf **March were strong contenders in the 1984 series with their 847 model, here at Dallas, driver Jim Crawford only just failing to take the championship back to Britain**

UNITED BREWERIES
INDIA

Above **March 817 driver Teo Fabi, driving for movie star Paul Newman, won more 1981 rounds than anyone else, had to be satisfied with runner-up position in the championship**

Right **One of the few US-built challengers not based on a Lola chassis, Al Holbert's CAC, here at Mosport in 1981, won several races, took top-three championship placing two years running**

Opposite,top **Danny Sullivan. later a leading light in among Champcar drivers, spent several seasons in Can-Am, here in an early Frissbee in the 1981 Mosport round**

Opposite,below **Also at Mosport in 1981, Geoff Brabham in the Team VDS Lola T530. He won one round in this car, and was more successful in the revised version known as the VDS-001**

was not as fast as his team-mate, but produced far more consistent results. His one first place, four seconds and three thirds gave him more points over the season than anyone but Ickx. For 1980 Newman reverted to Lolas, EFR finishing the championship in fourth place but English F2 driver Stephen South, driving the other car, suffering a career-ending crash at Trois Rivieres.

Another US project was run by Carl Hogan, who had been a regular entrant in North American F5000 racing, and one-time partner of Carl Haas. Hogan started his Can-Am association entering Lolas, in one of which Al Holbert was third in the 1978 series. The following year he entered his own Lola development, the Hogan HR001, which gave Holbert another top-six placing on the points table. It was in a Hogan that year that future champ Geoff Brabham got his first taste of Can-Am.

GOODYEAR
asi
4
GB-2
asi
4

BRABHAM
3
VALVOLINE
BOSCH

There were other British contenders, too, apart from Lola. The rule which allowed for racing engines under 3000cc, meant the Ford-Cosworth DFV engine, which had been dominating Grand Prix racing since 1967, could be used, and several entrants took advantage of the situation. One of the first was Alain de Cadenet, who entered a Mirage-DFV in a few 1978 races, with Howden Ganley also taking a turn at the wheel, but neither met with much success. For the 1982 series however. the Ensign company rebodied one of its N180 F1 cars and failed GP driver Jim Crawford managed two top-three placings with it, for a top six finish in the standings. Further development made the Ensign more competitive in 1983, and Crawford won two rounds of the championship which, backed by three second places, left him runner-up in the six-race series.

**By 1984 the big teams had mostly turned their backs on the revived Can-Am, leaving the locals to play among themselves. Best were the Frissbees, such as Horst Kroll's, here at Dallas**

The other British contender was March which, like Lola, had been a participant in the 'real' Can-Am series. Paul Newman's team ran the works effort in 1981, with works March F2 driver Teo Fabi as the star driver. In spite of a heavy mid-season crash at Elkhart Lake, the Italian won four rounds – more than anyone else that year – to place second on points. A second 817 was run first for veteran Al Unser and then for Bobby Rahal, but without the same success. The following year Danny Sullivan, a Can-Am regular over the previous two or three seasons after a successful European career in junior single-seater formulae, took his Newman March 827 to one first place, three seconds and two thirds, to end the series in third place.

March gave the 1983 series a miss but were back in 1984 when Jim Crawford won three of the ten rounds and was second in six others, though that was good enough only to give the 847 second place in the championship.

With March following Lola and VDS – not to mention the big entrants such as Hall, Haas, Hogan and Newman – to other fields, the final years of the Can-Am became the preserve of small US constructors. After his season with the Hogan in 1979, Al Holbert campaigned a car called the CAC over the next two seasons, winning five races and taking four second places, for second and third places in the respective points standings.

The last home product to score in the series was Frissbee. John Morton raced the first Lola-based car in 1980, and in the same year Al Unser won one round. Danny Sullivan repeated this effort in 1981, and then the following year 20-year-old Al Unser Jr won four races and was second in three others to give Frissbee their first Can-Am championship win. Jacques Villeneuve, brother of Gilles (and uncle of the present-day F1 driver) then won three of the six 1983 races in a similar car and was second in two of the others and so became the first Canadian to win the title.

But by now the second Can-Am series was on its last legs. The media, the sponsors and the public showed little interest in it for, apart from one or two top teams, there had never been much depth to the fields.

As grids continued to shrink each year, the 2-litre cars, originally admitted to make up numbers, were able to score more and more placings. US-resident Swede Bertil Roos, a former star of Super Vee and Formula Atlantic, was able to finish fourth overall in the 1982 championship with a small car and third the following year. By 1984, when grids were down to about a dozen cars, the 2-litre cars took even more placings.

But the series continued to limp on for two more years. Five races were held in 1985 and four in 1986, but the series was now little more than just another American club championship. Veteran Horst Kroll had finished fifth in the 1983 championship and third the following year, and in 1985 finally succeeded in winning one round. But the champion, with victories in three races, was series newcomer Rick Miaskiewicz with another Frissbee. Each of the 1986 races produced a different winner, but it was Kroll who took the title.

For what it was worth.

CAN-AM

# RESULTS

## CANADIAN-AMERICAN CHALLENGE CUP

### 1966

St Jovite 11 September

| | |
|---|---|
| 1 John Surtees | Lola-Chevrolet T70 Mk2 |
| 2 Bruce McLaren | McLaren-Chevrolet M1B |
| 3 Chris Amon | McLaren-Chevrolet M1B |

Bridgehampton 18 September

| | |
|---|---|
| 1 Dan Gurney | Lola-Ford T70 Mk2 |
| 2 Chris Amon | McLaren-Chevrolet M1B |
| 3 Bruce McLaren | McLaren-Chevrolet M1B |

Mosport 24 September

| | |
|---|---|
| 1 Mark Donohue | Lola-Chevrolet T70 |
| 2 Phil Hill | Chaparral-Chevrolet 2E |
| 3 Chuck Parsons | McLaren-Chevrolet M1B |

Laguna Seca 16 October

| | |
|---|---|
| 1 Phil Hill | Chaparral-Chevrolet 2E |
| 2 Jim Hall | Chaparral-Chevrolet 2E |
| 3 Bruce McLaren | McLaren-Chevrolet M1B |

*Los Angeles Times* Grand Prix, Riverside 30 October

| | |
|---|---|
| 1 John Surtees | Lola-Chevrolet T70 Mk2 |
| 2 Jim Hall | Chaparral-Chevrolet 2E |
| 3 Graham Hill | Lola-Chevrolet T70 Mk2 |

Stardust GP, Las Vegas, 13 November

| | |
|---|---|
| 1 John Surtees | Lola-Chevrolet T70 Mk2 |
| 2 Bruce McLaren | McLaren-Chevrolet M1B |
| 3 Mark Donohue | Lola-Chevrolet T70 |

FINAL POINTS

| | |
|---|---|
| 1 John Surtees (Lola) | 27 |
| 2 Mark Donohue (Lola) | 21 |
| 3 Bruce McLaren (McLaren) | 20 |
| 4 Phil Hill (Chaparral) | 18 |
| 5 Jim Hall (Chaparral) | 12 |
| 6 Chris Amon (McLaren) | 10 |

### 1967

Elkhart Lake 3 September

| | |
|---|---|
| 1 Denis Hulme | McLaren-Chevrolet M6A |
| 2 Mark Donohue | Lola-Chevrolet T70 Mk3B |
| 3 John Surtees | Lola-Chevrolet T70 Mk3B |

Bridgehampton 17 September

| | |
|---|---|
| 1 Denis Hulme | McLaren-Chevrolet M6A |
| 2 Bruce McLaren | McLaren-Chevrolet M6A |
| 3 George Follmer | Lola-Chevrolet T70 Mk3 |

Mosport 23 September

| | |
|---|---|
| 1 Denis Hulme | McLaren-Chevrolet M6A |
| 2 Bruce McLaren | McLaren-Chevrolet M6A |
| 3 Mike Spence | McLaren-Chevrolet M1B |

Laguna Seca 15 October

| | |
|---|---|
| 1 Bruce McLaren | McLaren-Chevrolet M6A |
| 2 Jim Hall | Chaparral-Chevrolet 2G |
| 3 George Follmer | Lola-Chevrolet T70 Mk3 |

Riverside 29 October

| | |
|---|---|
| 1 Bruce McLaren | McLaren-Chevrolet M6A |
| 2 Jim Hall | Chaparral-Chevrolet 2G |
| 3 Mark Donohue | Lola-Chevrolet T70 Mk3B |

Stardust GP, Las Vegas 12 November

| | |
|---|---|
| 1 John Surtees | Lola-Chevrolet T70 Mk2 |
| 2 Mark Donohue | Lola-Chevrolet T70 Mk3B |
| 3 Mike Spence | McLaren-Chevrolet M1B |

FINAL POINTS

| | |
|---|---|
| 1 Bruce McLaren (McLaren) | 30 |
| 2 Denis Hulme (McLaren) | 27 |
| 3 John Surtees (Lola) | 16 |
| 3 Mark Donohue (Lola) | 16 |
| 5 Jim Hall (Chaparral) | 15 |
| 6 George Follmer (Lola) | 10 |
| 6 Mike Spence (McLaren) | 10 |

### 1968

Road America 1 September

| | |
|---|---|
| 1 Denis Hulme | McLaren-Chevrolet M8A |
| 2 Bruce McLaren | McLaren-Chevrolet M8A |
| 3 Mark Donohue | McLaren-Chevrolet M6B |

Bridgehampton 15 September

| | |
|---|---|
| 1 Mark Donohue | McLaren-Chevrolet M6B |
| 2 Jim Hall | Chaparral-Chevrolet 2G |
| 3 Lothar Motschenbacher | McLaren-Ford M6B |

Edmonton 29 September

| | |
|---|---|
| 1 Denis Hulme | McLaren-Chevrolet M8A |
| 2 Bruce McLaren | McLaren-Chevrolet M8A |
| 3 Mark Donohue | McLaren-Chevrolet M6B |

Laguna Seca 13 October

| | |
|---|---|
| 1 John Cannon | McLaren-Chevrolet M1B |
| 2 Denis Hulme | McLaren-Chevrolet M8A |
| 3 George Eaton | McLaren-Chevrolet M1C |

LA Times Grand Prix, Riverside, 27 October

| | |
|---|---|
| 1 Bruce McLaren | McLaren-Chevrolet M8A |
| 2 Mark Donohue | McLaren-Chevrolet M6B |
| 3 Jim Hall | Chaparral-Chevrolet ?2H |

Las Vegas 10 November

| | |
|---|---|
| 1 Denis Hulme | McLaren-Chevrolet M8A |
| 2 George Follmer | Lola-Ford T70 |
| 3 Jerry Titus | McLaren-Chevrolet M6B |

FINAL POINTS

| | |
|---|---|
| 1 Denis Hulme (McLaren) | 35 |
| 2 Bruce McLaren (McLaren) | 24 |
| 3 Mark Donohue (McLaren) | 23 |
| 4 Jim Hall (Chaparral) | 12 |
| 5 L Motschenbacher (McLaren) | 11 |
| 6 John Cannon (McLaren) | 10 |

### 1969

Mosport 1 June

| | |
|---|---|
| 1 Bruce McLaren | McLaren-Chevrolet M8B |
| 2 Denis Hulme | McLaren-Chevrolet M8B |
| 2 John Surtees | McLaren-Chevrolet M12 |

St Jovite race 15 June

| | |
|---|---|
| 1 Denis Hulme | McLaren-Chevrolet M8B |
| 2 Bruce McLaren | McLaren-Chevrolet M8B |
| 3 Chuck Parsons | Lola-Chevrolet T163 |

Watkins Glen 13 July

| | |
|---|---|
| 1 Bruce McLaren | McLaren-Chevrolet M8B |
| 2 Denis Hulme | McLaren-Chevrolet M8B |
| 3 Chris Amon | Ferrari 612 |

Edmonton 27 July

| | |
|---|---|
| 1 Denis Hulme | McLaren-Chevrolet M8B |
| 2 Chris Amon | Ferrari 612 |
| 3 George Eaton | McLaren-Chevrolet M12 |

Mid-Ohio 17 August

| | |
|---|---|
| 1 Denis Hulme | McLaren-Chevrolet M8B |
| 2 Bruce McLaren | McLaren-Chevrolet M8B |
| 3 Chris Amon | Ferrari 612 |

Elkhart Lake 31 August

| | |
|---|---|
| 1 Bruce McLaren | McLaren-Chevrolet M8B |
| 2 Denis Hulme | McLaren-Chevrolet M8B |
| 3 Chuck Parsons | Lola-Chevrolet T163 |

Bridgehampton 14 September

| | |
|---|---|
| 1 Denis Hulme | McLaren-Chevrolet M8B |
| 2 Bruce McLaren | McLaren-Chevrolet M8B |
| 3 Jo Siffert | Porsche 917PA |

Michigan 28 September

| | |
|---|---|
| 1 Bruce McLaren | McLaren-Chevrolet M8B |
| 2 Denis Hulme | McLaren-Chevrolet M8B |
| 3 Dan Gurney | 'McLeagle'-Chevrolet (or M8B?) |

Laguna Seca 12 October

| | |
|---|---|
| 1 Bruce McLaren | McLaren-Chevrolet M8B |
| 2 Denis Hulme | McLaren-Chevrolet M8B |
| 3 Chuck Parsons | Lola-Chevrolet T163 |

Riverside 26 October

| | |
|---|---|
| 1 Denis Hulme | McLaren-Chevrolet M8B |
| 2 Chuck Parsons | Lola-Chevrolet T163 |
| 3 Mario Andretti | McLaren-Ford M6B |

College Station, Texas, 9 November

| | |
|---|---|
| 1 Bruce McLaren | McLaren-Chevrolet M8B |
| 2 George Eaton | McLaren-Chevrolet M12 |
| 3 Jack Brabham | AMR CA/Ford |

FINAL POINTS

| | |
|---|---|
| 1 Bruce McLaren (McLaren) | 165 |
| 2 Denis Hulme (McLaren) | 160 |
| 3 Chuck Parsons (Lola) | 81 |
| 4 Jo Siffert (Porsche) | 56 |
| 5 George Eaton (McLaren) | 50 |
| 6 Chris Amon (Ferrari) | 39 |

### 1970

Mosport 14 June

| | |
|---|---|
| 1 Dan Gurney | McLaren-Chevrolet M8D |
| 2 Jackie Oliver | Ti22-Chevrolet |
| 3 Denis Hulme | McLaren-Chevrolet M8D |

St Jovite 28 June

| | |
|---|---|
| 1 Dan Gurney | McLaren-Chevrolet M8D |
| 2 Lothar Motschenbacher | McLaren-Chevrolet M8B |
| 3 George Eaton | BRM-Chevrolet P154 |

Watkins Glen 12 July

| | |
|---|---|
| 1 Denis Hulme | McLaren-Chevrolet M8D |
| 2 Jo Siffert | Porsche 917 |
| 3 Dick Attwood | Porsche 917 |

Edmonton 26 July

| | |
|---|---|
| 1 Denis Hulme | McLaren-Chevrolet M8D |
| 2 Peter Gethin | McLaren-Chevrolet M8D |
| 2 Lothar Motschenbacher | McLaren-Chevrolet M8B |

Mid-Ohio 23 August

| | |
|---|---|
| 1 Denis Hulme | McLaren-Chevrolet M8D |
| 2 Peter Revson | Lola-Chevrolet T220 |
| 3 Lothar Motschenbacher | McLaren-Chevrolet M8B |

Road America 30 August

| | |
|---|---|
| 1 Peter Gethin | McLaren-Chevrolet M8D |
| 2 Bob Bondurant | Lola-Chevrolet T160 |
| 3 Dave Causey | Lola-Chevrolet T163 |

Road Atlanta 13 September

| | |
|---|---|
| 1 Tony Dean | Porsche 908 |
| 2 Dave Causey | Lola-Chevrolet T163 |
| 3 Lothar Motschenbacher | McLaren-Chevrolet M12 |

Donnybrooke 27 September

| | |
|---|---|
| 1 Denis Hulme | McLaren-Chevrolet M8D |
| 2 Peter Gethin | McLaren-Chevrolet M8D |
| 3 Peter Revson | Lola-Chevrolet T220 |

Laguna Seca 18 October

| | |
|---|---|
| 1 Denis Hulme | McLaren-Chevrolet M8D |
| 2 Jackie Oliver | Ti22-Chevrolet Mk2 |
| 3 Peter Revson | Lola-Chevrolet T220 |

Riverside 1 November

| | |
|---|---|
| 1 Denis Hulme | McLaren-Chevrolet M8D |
| 2 Jackie Oliver | Ti22-Chevrolet Mk2 |
| 3 Pedro Rodriguez | BRM-Chevrolet P154 |

FINAL POINTS

| | |
|---|---|
| 1 Denis Hulme (McLaren) | 132 |
| 2 L Motschenbacher (McLaren) | 65 |
| 3 Peter Gethin (McLaren) | 56 |
| 4 Dave Causey (Lola) | 56 |
| 5 Jackie Oliver (Ti22) | 45 |
| 6 Tony Dean (Porsche) | 44 |

## 1971

Mosport 13 June

| | |
|---|---|
| 1 Denis Hulme | McLaren-Chevrolet M8F |
| 2 Peter Revson | McLaren-Chevrolet M8F |
| 3 Lothar Motschenbacher | McLaren-Chevrolet M8D |

St Jovite 27 June

| | |
|---|---|
| 1 Jackie Stewart | Lola-Chevrolet T260 |
| 2 Denis Hulme | McLaren-Chevrolet M8F |
| 3 Peter Revson | McLaren-Chevrolet M8F |

Road Atlanta 11 July

| | |
|---|---|
| 1 Peter Revson | McLaren-Chevrolet M8F |
| 2 Denis Hulme | McLaren-Chevrolet M8F |
| 3 Lothar Motschenbacher | McLaren-Chevrolet M8D |

Watkins Glen 25 July

| | |
|---|---|
| 1 Peter Revson | McLaren-Chevrolet M8F |
| 2 Denis Hulme | McLaren-Chevrolet M8F |
| 3 Jo Siffert | Porsche 917/10 |

Mid-Ohio 22 August

| | |
|---|---|
| 1 Jackie Stewart | Lola-Chevrolet T260 |
| 2 Jo Siffert | Porsche 917/10 |
| 3 Tony Adamowicz | McLaren-Chevrolet M8B |

Elkhart Lake 29 August

| | |
|---|---|
| 1 Peter Revson | McLaren-Chevrolet M8F |
| 2 Jo Siffert | Porsche 917/10 |
| 3 Vic Elford | McLaren-Chevrolet M8E |

Donnybrooke 12 September

| | |
|---|---|
| 1 Peter Revson | McLaren-Chevrolet M8F |
| 2 Denis Hulme | McLaren-Chevrolet M8F |
| 3 Gregg Young | McLaren-Chevrolet M8E/D |

Edmonton 26 September

| | |
|---|---|
| 1 Denis Hulme | McLaren-Chevrolet M8F |
| 2 Jackie Stewart | Lola-Chevrolet T260 |
| 3 Jackie Oliver | Shadow-Chevrolet Mk2 |

Laguna Seca 17 October

| | |
|---|---|
| 1 Peter Revson | McLaren-Chevrolet M8F |
| 2 Jackie Stewart | Lola-Chevrolet T260 |
| 3 Denis Hulme | McLaren-Chevrolet M8F |

Riverside 31 October

| | |
|---|---|
| 1 Denis Hulme | McLaren-Chevrolet M8F |
| 2 Peter Revson | McLaren-Chevrolet M8F |
| 3 Howden Ganley | BRM-Chevrolet P167 |

FINAL POINTS

| | |
|---|---|
| 1 Peter Revson (McLaren) | 142 |
| 2 Denis Hulme (McLaren) | 132 |
| 3 Jackie Stewart (Lola) | 76 |
| 4 Jo Siffert (Porsche) | 68 |
| 5 L Motschenbacher (McLaren) | 52 |
| 6 Milt Minter (Porsche) | 37 |

## 1972

Mosport 11 Junes

| | |
|---|---|
| 1 Denis Hulme | McLaren-Chevrolet M20 |
| 2 Mark Donohue | Porsche 917/10T |
| 3 Peter Revson | McLaren-Chevrolet M20 |

Road Atlanta 9 July

| | |
|---|---|
| 1 George Follmer | Porsche 917/10T |
| 2 Gregg Young | McLaren-Chevrolet M8F |
| 3 Milt Minter | Porsche 917/10 |

Watkins Glen 23 July

| | |
|---|---|
| 1 Denis Hulme | McLaren-Chevrolet M20 |
| 2 Peter Revson | McLaren-Chevrolet M20 |
| 3 François Cevert | McLaren-Chevrolet M8F |

Mid-Ohio 6 August

| | |
|---|---|
| 1 George Follmer | Porsche 917/10T |
| 2 Jackie Oliver | Shadow-Chevrolet Mk3 |
| 3 Milt Minter | Porsche 917/10 |

Road America 27 August

| | |
|---|---|
| 1 George Follmer | Porsche 917/10T |
| 2 François Cevert | McLaren-Chevrolet M8F |
| 3 Peter Gregg | Porsche 917/10T |

Donnybrooke 17 September

| | |
|---|---|
| 1 François Cevert | McLaren-Chevrolet M8F |
| 2 Milt Minter | Porsche 917/10 |
| 3 Jackie Oliver | Shadow-Chevrolet Mk3 |

Edmonton 1 October

| | |
|---|---|
| 1 Mark Donohue | Porsche 917/10T |
| 2 Denis Hulme | McLaren-Chevrolet M20 |
| 3 George Follmer | Porsche 917/10T |

Laguna Seca 15 October

| | |
|---|---|
| 1 George Follmer | Porsche 917/10T |
| 2 Mark Donohue | Porsche 917/10T |
| 3 François Cevert | McLaren-Chevrolet M8F |

Riverside 29 October

| | |
|---|---|
| 1 George Follmer | Porsche 917/10T |
| 2 Peter Revson | McLaren-Chevrolet M20 |
| 3 Mark Donohue | Porsche 917/10T |

FINAL POINTS

| | |
|---|---|
| 1 George Follmer (Porsche) | 130 |
| 2 Denis Hulme (McLaren) | 65 |
| 3 Milt Minter (Porsche) | 65 |
| 4 Mark Donohue (Porsche) | 62 |
| 5 François Cevert (McLaren) | 59 |
| 6 Peter Revson (McLaren) | 48 |

## 1973

Mosport 10 June

| | |
|---|---|
| 1 Charlie Kemp | Porsche 917/10T |
| 2 Hans Wiedmer | Porsche 917/10T |
| 3 Bob Nagel | Lola-Chevrolet T260 |

Road Atlanta 7 July

| | |
|---|---|
| 1 George Follmer | Porsche 917/10T |
| 2 Mark Donohue | Porsche 917/30 |
| 3 Jody Scheckter | Porsche 917/10T |

Watkins Glen 22 July

| | |
|---|---|
| 1 Mark Donohue | Porsche 917/30 |
| 2 David Hobbs | McLaren-Chevrolet M20 |
| 3 Jody Scheckter | Porsche 917/10T |

Mid-Ohio 12 August

| | |
|---|---|
| 1 Mark Donohue | Porsche 917/30 |
| 2 George Follmer | Porsche 917/10T |
| 3 Hurley Haywood | Porsche 917/10T |

Road America 26 August

| | |
|---|---|
| 1 Mark Donohue | Porsche 917/30 |
| 3 Jody Scheckter | Porsche 917/10T |
| 3 George Follmer | Porsche 917/10T |

Edmonton 16 September

| | |
|---|---|
| 1 Mark Donohue | Porsche 917/30 |
| 2 George Follmer | Porsche 917/10T |
| 3 Jackie Oliver | Shadow-Chevrolet DN2 |

Laguna Seca 14 October

| | |
|---|---|
| 1 Mark Donohue | Porsche 917/30 |
| 2 Jackie Oliver | Shadow-Chevrolet DN2 |
| 3 Hurley Haywood | Porsche 917/10T |

Riverside 28 October

| | |
|---|---|
| 1 Mark Donohue | Porsche 917/30 |
| 2 Hurley Haywood | Porsche 917/10T |
| 3 Charlie Kemp | Porsche 917/10T |

FINAL POINTS

| | |
|---|---|
| 1 Mark Donohue (Porsche) | 139 |
| 2 George Follmer (Porsche) | 62 |
| 3 Hurley Haywood (Porsche) | 47 |
| 4 Charlie Kemp (Porsche) | 45 |
| 5 Bob Nagel (Lola) | 44 |
| 6 Jody Scheckter (Porsche) | 39 |

## 1974

Mosport 16 June

| | |
|---|---|
| 1 Jackie Oliver | Shadow-Chevrolet DN4 |
| 2 George Follmer | Shadow-Chevrolet DN4 |
| 3 Scooter Patrick | McLaren-Chevrolet M20 |

Road Atlanta 7 July

| | |
|---|---|
| 1 Jackie Oliver | Shadow-Chevrolet DN4 |
| 2 George Follmer | Shadow-Chevrolet DN4 |
| 3 Lothar Motschenbacher | McLaren-Chevrolet M8F |

Watkins Glen 14 July

| | |
|---|---|
| 1 Jackie Oliver | Shadow-Chevrolet DN4 |
| 2 George Follmer | Shadow-Chevrolet DN4 |
| 3 Scooter Patrick | McLaren-Chevrolet M20 |

Mid-Ohio 11 August

| | |
|---|---|
| 1 Jackie Oliver | Shadow-Chevrolet DN4 |
| 2 Brian Redman | Porsche 917/30 |
| 3 Hurley Haywood | Porsche 917/10 |

Road America 25 August

| | |
|---|---|
| 1 Scooter Patrick | McLaren-Chevrolet M20 |
| 2 John Cordts | McLaren-Chevrolet M8F |
| 3 John Gunn | Lola-Chevrolet T260 |

FINAL POINTS

| | |
|---|---|
| 1 Jackie Oliver (Shadow) | 82 |
| 2 George Follmer (Shadow) | 45 |
| 3 Scooter Patrick (McLaren) | 44 |
| 4 Bob Nagel (Lola) | 40 |
| 5 John Gunn (Lola) | 23 |
| 6 L Motschenbacher (McLaren) | 21 |

CAN-AM

# Colour PROFILES

Pages 87 and 88

Chaparral was just one of many US challengers in the early Can-Am series, but the only really serious one. Team owner Jim Hall was already renowned for innovation, especially in the aerodynamics field, when the series was inaugurated in 1966. His 2E took one step further the experiments he had conducted on earlier cars, with tail-mounted "flipper" aerofoil replaced by one high above the car and bolted directly to the suspension. So successful was this development that within two years it had revolutionised Formula One design.

McLaren were strong contenders with their first spaceframe cars in 1966, then won the series with their monocoque M6A in 1967. The M8A ran a bigger engine which meant more fuel, so the whole car, designed by Jo Marquart and Gordon Coppuck, was bigger than the M6A. But careful attention to weight-saving measures meant it was no heavier – and had 20% more horsepower to propel it. The engine suffered somewhat from teething troubles, but the M8A was as dominant in the 1968 Can-Am series as the M6A had been the year before.

More weight was pared from the design for the 1969 McLaren M8B. and the suspension layout was improved. But the biggest change was the use of a suspension-mounted rear aerofoil. By then every F1 team had copied the Chaparral 2E idea, but a series of breakages had led to their being banned from GP cars in 1968. Only aerofoils mounted on the bodywork could be used. This restriction took a year to filter down to Can-Am, and McLaren enjoyed another successful season with the high-winged M8B.

Lola won the first Can-Am series with the T70 in 1966, but then suffered a lean spell, with successive replacements being good for solid placings rather than contenders for outright victory. Until the T260 of 1971. This model departed from accepted Can-Am design practice of the time by using a blunt nose to reduce drag and a relatively small rear wing. Lola chief Eric Broadley's design coupled with the driving skills of Jackie Stewart resulted in a winner, capable of beating the McLaren juggernaut fair and square, though more reliability would have helped.

Porsche tried year after year to make the same sort of mark in the Can-Am series as they had with the dominant 917 in world endurance championship events such as the Le Mans 24hr and the Nürburgring 1000km. At the same time turbocharged engines had begun to take over Indycar racing, and it was Porsche who first successfully applied this technology to a Can-Am car engine, with their 917/10 in 1972. Such was the power available to the drivers of these cars that the McLarens were finally brought down, and Follmer won the title for entrant Roger Penske.

Having found the right formula, Porsche set about improving on it. Penske's driver/engineer Mark Donohue went to Germany to help the factory perfect the design, and the 917/30 which resulted was better streamlined. More significantly, power increased from 900bhp to 1100 bhp, and that was more than even the Indycars were then producing, and certainly far above F1 horsepower figures. The Porsche/Penske/Donohue effort with the 917/30 was quite unbeatable in 1973, leading the series organisers to revise their rules at the end of the year.

In terms of native American Can-Am challengers, Shadow took up where Chaparral left off. Always competitive, the black cars made their mark early, and returned to each year's series in a stronger position. By the time of the DN4 model of 1974, the operation had moved its base to England, and had given up fruitless experiments with turbocharged engines. The other factor in their 1974 success was that Porsche had now withdrawn from the series because of the rule changes, and with McLaren, Lola and Chaparral long gone, there was no longer any serious team competition.